Climate Change Resilience in the Urban Environment

Climate Change Resilience in the Urban Environment

Tristan Kershaw
University of Bath

IOP Publishing, Bristol, UK

Media content for this book is available from https://doi.org/10.1088/978-0-7503-1197-7.

ISBN 978-0-7503-1197-7 (ebook)
ISBN 978-0-7503-1198-4 (print)
ISBN 978-0-7503-1199-1 (mobi)

DOI 10.1088/978-0-7503-1197-7

Version: 20171201

IOP Expanding Physics
ISSN 2053-2563 (online)
ISSN 2054-7315 (print)

British Library Cataloguing-in-Publication Data: A catalogue record for this book is available from the British Library.

Published by IOP Publishing, wholly owned by The Institute of Physics, London

IOP Publishing, Temple Circus, Temple Way, Bristol, BS1 6HG, UK

US Office: IOP Publishing, Inc., 190 North Independence Mall West, Suite 601, Philadelphia, PA 19106, USA

This book is dedicated to my family; my lovely wife Caroline and our son Jasper, whose arrival during the writing of the book proved a major (but welcome) distraction from writing and whose little fingers were constantly reaching for my keyboard, trying to edit my work. So if you find any mistakes they are probably his fault - honest.

Contents

Preface **ix**

Author biography **xi**

1 Climate change and its impacts **1-1**

1.1 The greenhouse effect 1-2
1.2 The historic climate signal 1-8
1.3 The anthropogenic greenhouse effect 1-14
1.4 Climate change projections 1-19
1.5 Climate change impacts 1-23
References 1-27

2 Water **2-1**

2.1 Sea level rise 2-2
2.2 Storm surges 2-6
2.3 Flooding 2-8
2.4 Flash flooding 2-14
2.5 Potential solutions 2-16
2.6 Conclusions 2-23

3 Temperatures **3-1**

3.1 Human physiology 3-6
3.2 Building physics and possible adaptations 3-15
3.3 Learning from other architectures 3-21
3.4 Summary 3-29
References 3-30

4 The urban heat island (UHI) **4-1**

4.1 Boundary layer creation 4-2
4.2 The energetic basis and UHI creation 4-5
4.3 Weather influence 4-9
4.4 Implications of the UHI on the built environment 4-11
4.5 Air quality in cities 4-14
4.6 Green and blue infrastructure 4-20
4.7 Thermal effects of green space 4-20
4.8 Green space implications for city planning 4-26

4.9 Green building envelopes 4-30
4.10 Thermal properties of blue space 4-31
4.11 Thermal effects of blue space 4-36
4.12 Urban planning for the UHI 4-40
References 4-44

5 Weather extremes 5-1

5.1 Heatwaves 5-1
5.2 Storms 5-5
Reference 5-13

6 Conclusions 6-1

6.1 Building resilience 6-2
6.2 Urban resilience 6-11

Preface

For the first time in history more than half of the world's population lives in urban areas, a number that was only 30% in 1960, and this figure is set to increase dramatically over the next few decades, reaching ~70% by 2050. This rapid growth in urban populations will occur mainly in the developing world, with some countries expected to experience up to a five-fold increase in urban populations by 2050. Increasing urbanisation brings with it a series of benefits, but also challenges. Increased urbanisation and urban density allows for more efficient use of materials and infrastructure, such as fewer roads, more efficient public transport, reduced distribution losses in energy networks, and the ability to have larger more-centralised public services, e.g. police, fire services, and healthcare. However, there are draw backs. An increasing urban area typically means a loss of vegetative cover in favour of hard impermeable man-made surfaces. This can increase the risk of flooding from intense rainfall, while the dense materials used to construct our urban areas store heat from the Sun to be radiated later, altering the local climate. Even the position and orientation of our buildings can have a negative effect, isolating the urban microclimate from the atmosphere above, trapping heat and pollutants at street level.

Human civilisation and current architectural practices have evolved over the past few thousand years during a period of relatively constant climate and predictable weather. Thus, our buildings are the result of the interaction between culture, geography, and meteorology. This has allowed for architectural design and the layout of urban areas to be tailored to the local climate. Underneath the influences of culture and social interaction, the architecture of a building has two main functions: to provide shelter from the elements, and to provide thermal comfort for its occupants. At every latitude, buildings transform sometimes harsh exterior environmental conditions into a comfortable interior environment. They achieve this feat through little more than careful design and the appropriate use of materials. However, as cities have begun to grow rapidly, these traits have been lost.

Globalisation has led to the prevalence of new materials and techniques, creating a divergence between the design and construction of buildings and their local cultural, social, ecological, and environmental context. Buildings are now built more for speed of construction and cost rather than to be resilient to the local climate, as cheap energy has meant that artificial heating and cooling can be used to compensate for a climate-insensitive design.

Climate change is altering the way our buildings will need to perform. Changing temperatures, the seasonality of rainfall, and increased instances of weather that are currently classed as extreme, mean that our buildings will have to cope with different environmental conditions than previously. There is evidence, however, that vernacular architecture can be integrated into modern buildings to promote resilience to environmental stresses. But in order to combat the impacts of climate change, we will need to consider not only the vernacular architectural details from our own culture and climate, but also those of other global locations that currently have a climate that may be similar to what we can expect in the future.

Figure 1. Growing cities and expanding urban areas can provide many socioeconomic benefits but also provide many climate-related challenges.

This book aims to consider not only what the impacts of climate change are on the built environment, but also what we can do to try and mitigate the impacts of climate change.

Author biography

Tristan Kershaw

Tristan Kershaw is currently a Lecturer in Climate Resilience at the University of Bath. Tristan graduated from the University of Exeter in 2004 with a Masters degree in Physics and went on to study for a PhD in low temperature solid state physics. After completing his PhD, he joined the Centre for Energy and the Environment also at the University of Exeter as a research fellow in climate change adaptation. Over the subsequent six years, he worked on a variety of 'building physics'-related research and consultancy projects, including the creation of probabilistic future weather years for the UK for the thermal modeling of buildings, as well as the modelling and adaptation of building designs for several exemplar buildings across the southwest region. In 2014, Tristan joined the University of Bath as a lecturer in the Department of Architecture and Civil Engineering, teaching both undergraduate and postgraduate engineers and architects on the topics of building physics, sustainability, climate change, and the dynamic modelling of building designs.

This book is based primarily upon Tristan's research and consultancy work, and as such it contains many case studies from, in, and around the southwest UK, along with observations of architecture and building use from further afield.

IOP Publishing

Tristan Kershaw

Chapter 1

Climate change and its impacts

Climate is a difficult concept for people to deal with, as generally we think in terms of short-term variations or weather, and our memory is drawn towards more extreme events such as heat waves, cold snaps, and storms. Climate, however, is defined as the long-term averages and ranges of different weather variables. Changes in climate typically take many thousands of years; hence, human civilisation has evolved during a period of relatively constant climate. This means that buildings, urban areas, and even human physiology are ill-adapted to relatively rapid changes in climate over several decades or centuries.

The world is experiencing a massive transition from rural to urban living. Urban areas occupy less than 2% of the Earth's land surface, but house over 50% of the world's population, a figure that was only 14% in 1900 and one which is estimated to increase to 60% by 2030. This rapid rise will mainly take place in developing countries. Asia and Africa will see the largest increases in urban population, which are expected to reach 64% and 56% urban population, respectively, by 2050. Countries such as Mali, Niger, Tanzania, Uganda, and Zambia are predicted to see greater than a five-fold increase in urban population by 2050. If the developing world continues to urbanise as expected, replacing sprawling urban areas with extremely dense ones, they will face a plethora of problems. Denser urban areas are generally considered to be more sustainable, requiring less land, infrastructure, and being resource efficient; and it is highly probable that future cites in currently developing countries will be at least as dense as those in the developed world. However, this comes at the cost of poorer air quality, reduced biodiversity, reduced flood resilience, higher air temperatures due to the urban heat island, and potentially poorer physical and mental health associated with reduced green (grass, trees, and vegetation) and blue space (reservoirs, lakes, rivers, canals etc).

Climate change will affect the performance of buildings, urban areas, and the comfort and health of a city's occupants. This chapter will cover the origins and scientific basis for climate change before presenting an overview of the impacts of

doi:10.1088/978-0-7503-1197-7ch1

climate change based upon the latest climate projections. Using these projections, the following chapters will examine the potential impacts of climate change on urban areas, and consider how to increase the resilience of our growing urban areas to a rapidly changing climate.

1.1 The greenhouse effect

The climate Earth experiences is a function of its distance from the Sun and the amount of radiation (energy) received. At any given time, the surface of the Earth must be in equilibrium with its surroundings; therefore, the Earth must be re-radiating an equal amount of energy to that received from the Sun. All objects at a temperature above absolute zero (0 K, −273 °C) emit radiation. The hotter the object, the shorter the mean wavelength (λ) of this radiation. The peak wavelength of the distribution (black-body curve) of radiation emitted, λ_{max} (in metres, m), is given by Wien's displacement law:

$$\lambda_{\text{max}} = \frac{2.9 \times 10^{-3}}{T}$$

where T (in Kelvin, K) is the temperature of the object. For the Sun, with a surface temperature of ~5800 K, the peak is at ~500 nm (500×10^{-9} m), which is in the visible light region. For the Earth, with its much lower temperature, the radiation is of a longer wavelength and within the infrared part of the electromagnetic spectrum.

The Sun is not a perfect black-body emitter, and not all the wavelengths of radiation given off reach the Earth's surface. Ozone (O_3) in the atmosphere absorbs and re-radiates ultra-violet (UV) light. Rayleigh scattering by dust particles in the upper atmosphere scatters blue light, which is why we view the sky as blue. Other wavelengths of light primarily in the near-infrared region are absorbed and re-radiated by other gases in the atmosphere such as oxygen (O_2) and water vapour (H_2O). These absorption windows are associated with different mechanisms of bending and stretching molecular bonds. The solar spectrum both at the top of the atmosphere and what reaches the surface of the Earth can be seen in figure 1.1.

Since we know the temperature of the Sun from its colour and spectral output and we also know the distance of the Earth from the Sun, we can calculate how much radiation the Earth receives, and hence the temperature the Earth should be to be in equilibrium, i.e. radiating an equivalent amount of energy to that it receives.

A hot object will radiate with an intensity I (W m^{-2}, Watts per square metre) given by

$$I = \sigma T^4$$

where σ is the Stefan–Boltzmann constant (5.67×10^{-8} W m^{-2} K^{-4}). The surface of the Sun is at around ~5777 K, and hence radiates 6.32×10^7 W m^{-2}. The solar radius is 6.96×10^8 m (696 000 km), and the surface area ($4\pi r^2$) of the Sun is 6.1×10^{18} m^2; therefore, the Sun must radiate with a luminosity of 3.85×10^{26} W. Assuming this radiation is emitted equally in all directions, we can estimate how

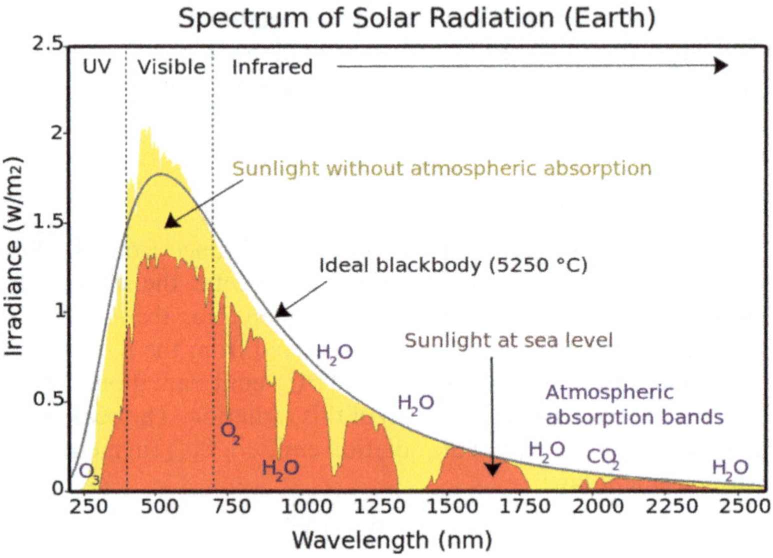

Figure 1.1. Solar spectrum at the top of the atmosphere and at sea level (This solar spectrum image has been obtained by the author from the Wikimedia website where it was made available under a CC BY-SA 3.0 licence. It is included within this article on that basis. It is attributed to Nick84.

much of this is intercepted by the Earth. If we imagine this radiation smeared across the inside of a sphere centred on the Sun and intersecting the Earth, as shown in figure 1.2, we can work out the fraction of the energy intercepted by the Earth from the ratio of the two surface areas, i.e. of the sphere and the Earth.

The Earth is at a distance of one Astronomical Unit (1 AU) from the Sun, which is equivalent to 1.496×10^{11} m (149.6 million km). Therefore, the surface area of the sphere intersecting the Earth is 2.81×10^{23} m^2. The Earth will project an area of πr^2 onto the surface of this sphere, which with a radius of 6.371×10^6 m (6371 km) equals 1.275×10^{14} m^2. Taking the ratio between the surface area of the sphere and the area the Earth projects on the surface of this sphere, we can calculate that the Earth receives only $1/4.5 \times 10^{10\text{th}}$ of the radiation from the Sun. Or in simpler terms, the Earth receives 1.74×10^{17} W from the Sun.

Not all the radiation received by the planet is absorbed. The Earth has an albedo of ~0.3, meaning ~30% is reflected back directly into space from the atmosphere and the ground. This albedo varies slightly according to the amount of cloud cover in the atmosphere and where it is situated. Thus, the planet absorbs approximately 1.22×10^{17} W, which when averaged over the whole surface of the Earth ($4\pi r^2$), is approximately ~239 W m^{-2}. The planet must be in radiative balance (equilibrium), re-radiating all radiation received; else the planet's temperature would continuously increase. Hence, working backwards using $I = \sigma T^4$, we can estimate the temperature of the Earth required to be in radiative balance. This suggests that the temperature of the Earth should be ~255 K, or −18 °C. While this value calculated is close to the measured temperature of the Moon, clearly, this is not true for the Earth. The average temperature of the Earth is closer to 15 °C (298 K). The reason for

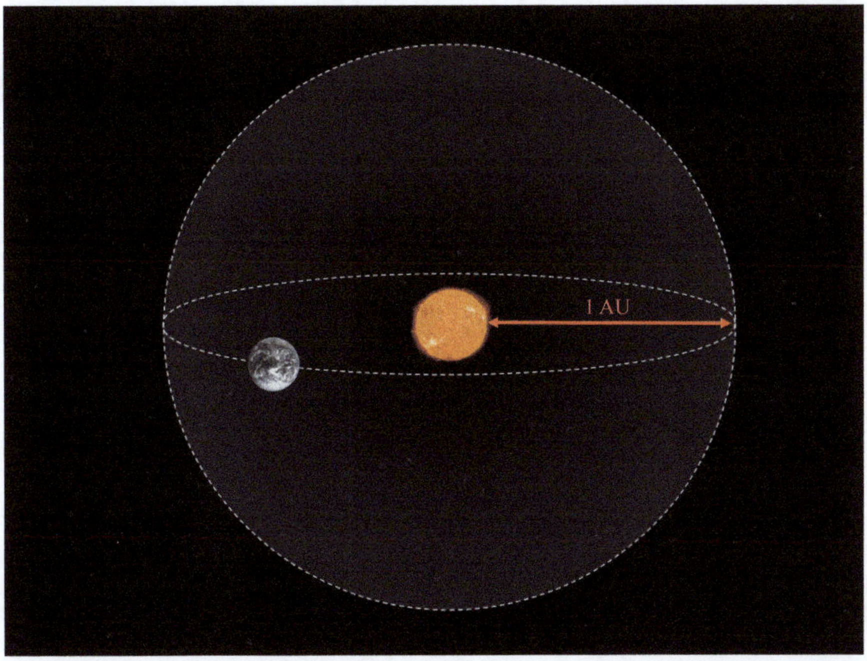

Figure 1.2. Consider a sphere of radius 1 AU centred on the Sun, with the Earth intersecting the surface.

this, which all scientists agree on, lies in the makeup and chemistry of the Earth's atmosphere.

This atmospheric chemistry, which is responsible for the ~33 °C disparity between the observed and calculated temperature of the Earth's surface, is termed the natural greenhouse effect. The basis of this effect lies in the relative absorption spectra of the different atmospheric gases. Figure 1.3 shows the Earth in the visible spectrum; the result is the iconic blue marble that we are all familiar with, indicating that the entire visible spectrum is able to pass unhindered through the atmosphere to the surface. Light from the Sun travels through space, through the atmosphere, hits the surface of the Earth (in the case of figure 1.3, a grain of sand in the Sahara), and is reflected back through the atmosphere of the Earth to the camera lens, unobstructed and without scattering. The fact that we can see the surface of the Earth from space in great detail tells us that we have not lost any directional information, and that light is not scattered, diverted, or absorbed during its passage through the atmosphere. The only scattering comes from clouds in the atmosphere; because they appear white, we can deduce that they scatter all visible light equally. The result is that of the solar radiation intercepting the Earth, around 70% is absorbed, contributing to heating the surface, while ~30% is reflected from the surface or from clouds in the atmosphere. The atmosphere itself is transparent. For comparison, figure 1.4 shows the Earth viewed at two wavelengths in the infrared spectrum. The left-hand image shows the Earth viewed at 6.7 μm (6.7 microns or 6.7×10^{-6} m), which is within a water vapour absorption band. We can see that we have lost all directional information, and light is being fully scattered (absorbed and

Figure 1.3. The Earth viewed from space in the visible light spectrum (courtesy of NASA).

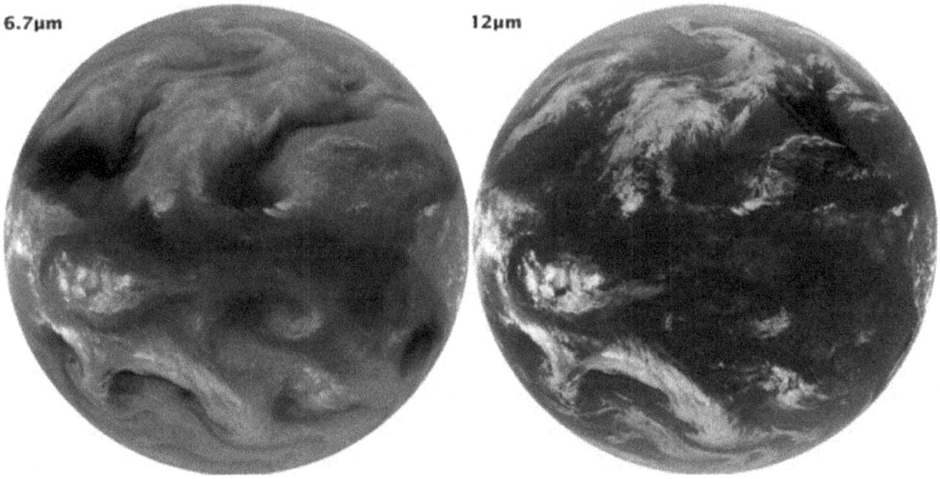

Figure 1.4. The Earth viewed from space in the infrared spectrum (courtesy of NASA).

re-radiated) within the atmosphere. The right-hand image shows the same image of the Earth but viewed at 12 μm, which while in the infrared spectrum is largely transmitted by the atmospheric greenhouse gases. Here we can see we have lost some directional information: the image is milky not crystal clear, but we can still make out the continents. This is the origin of the natural greenhouse effect and the additional warming the Earth receives as a result. Visible light is able to pass relatively unhindered through the atmosphere to the surface of the planet, thus warming the surface. The warm surface of the planet radiates according to its temperature in the infrared spectrum. However, these wavelengths of light find it much harder to pass through the atmosphere, and are reflected back to the surface, or are absorbed and re-radiated by gases in the atmosphere. The absorption (and transmission) spectra of the atmosphere as a whole and the most important greenhouse gases are shown in figure 1.5.

The absorption spectrum arises from the absorption of light and its conversion into mechanical energy; in this case, vibrations of molecular bonds and the motion of constituent atoms. This energy is transferred to the surrounding gases (principally nitrogen and oxygen, which are not greenhouse gases), thus leading to a warming of the atmosphere.

Although we have to have a radiative balance at the top of the atmosphere for the planet to be in equilibrium, for the surface and the rest of the atmosphere we need a more complex approach. We need to take account of the absorption, reflection, heat

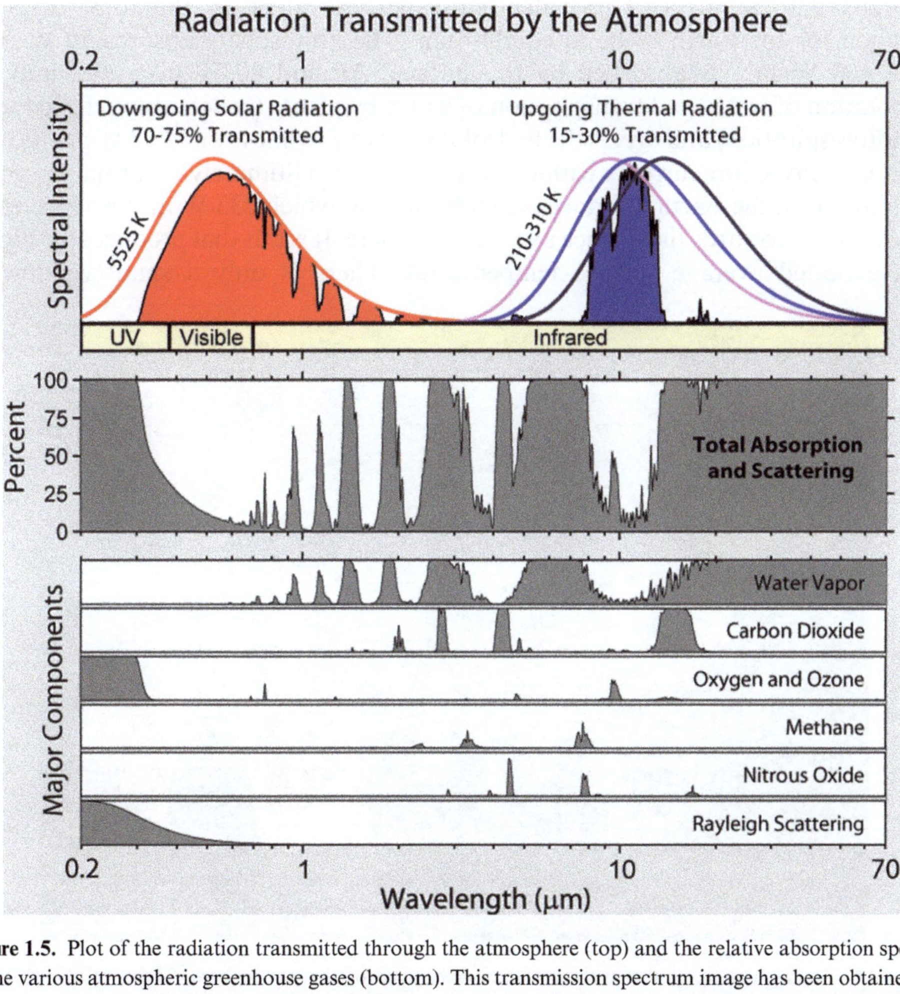

Figure 1.5. Plot of the radiation transmitted through the atmosphere (top) and the relative absorption spectra of the various atmospheric greenhouse gases (bottom). This transmission spectrum image has been obtained by the author from the Wikimedia website where it was made available under a CC BY-SA 3.0 licence. It is included within this article on that basis. It is attributed to Global Warming Art.

transfer, and convection within the atmosphere. As we can see from figure 1.5, different atmospheric gases absorb different wavelengths of light, and these gases must re-radiate all the energy they absorb; however, unlike the surface of the Earth, they are able to radiate in all directions, not just upwards. Hence, some of this radiation will be re-radiated back towards the Earth's surface and contribute to additional heating. This is this natural greenhouse effect that produces the elevated surface temperatures and results in the average temperature of the Earth being 15 °C rather than −18 °C. We can see from figure 1.5 that water vapour is the largest contributor to atmospheric absorption. Additionally, water vapour is responsible for approximately 22 °C of the 33 °C additional warming experienced as a result of the natural greenhouse effect.

Figure 1.6 provides an illustration of the typical magnitudes of the radiative flux experienced through the atmosphere. Of the 341 W m^{-2} that hits the top of the atmosphere, around 30% is reflected back out to space (net downwards radiation absorbed 239 W m^{-2}). This value must be equivalent to the total outgoing radiation for the Earth to be in equilibrium. The atmosphere absorbs 79 W m^{-2}, while 160 W m^{-2} is absorbed by the surface. Around 80 W m^{-2} goes into the evaporation of water and transpiration of water by plants (collectively referred to as evapotranspiration) and drives the hydrological cycle, while around 17 W m^{-2} is used to set up convection currents within the atmosphere. Ultimately, the total up-going radiation from the Earth's surface is 396 W m^{-2}, of which 333 W m^{-2} is reflected or radiated back towards the surface by the atmosphere. It is this that produces the higher than expected average surface temperatures. There is only a small window of

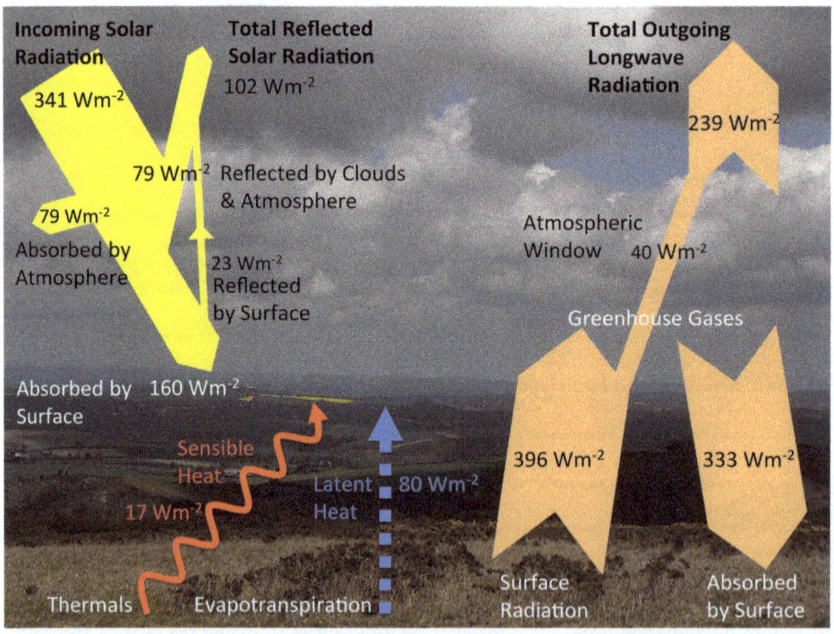

Figure 1.6. Representation of the average radiation balance of the atmosphere. These values will vary according to season, latitude, and with astronomical mechanics.

wavelengths where heat (infrared radiation) can pass through the atmosphere unimpeded, accounting for only 40 W m^{-2}. The remainder of this radiation warms the atmosphere, as do the sensible and latent heat contributions.

1.2 The historic climate signal

While human civilisation has arisen during a period of relatively constant climate, the Earth has seen many different climates over its 4.5 billion year history. Through examination of the fossil record and deep ice cores, we can draw a picture of what life on Earth was like at different time periods.

There are stable isotopes of oxygen, oxygen-16 (^{16}O), which contains 8 protons and 8 neutrons, and the less common oxygen-18 (^{18}O), containing 8 protons and 10 neutrons. In the paleosciences the ratio of ^{18}O : ^{16}O ($\delta^{18}O$) found in corals, fossils, and ice cores can be used as a proxy for temperature. This arises from the differential rates at which water molecules containing these isotopes evaporate or condense. When water vapour condenses, the heavier water molecules containing ^{18}O atoms condense and precipitate first. Thus, there is a preferential evaporation of ^{16}O from seawater, and hence fresh water precipitation is ^{16}O-enriched, leading to a gradient in the $\delta^{18}O$ with latitude. Ocean surfaces contain greater amounts of ^{18}O around the tropics where there is increased evaporation and reduced amounts of ^{18}O at the mid-latitudes where there is more rain. Additionally, the amount of ^{18}O present in water vapour is greater at the tropics than closer to the poles, due to higher temperatures and greater evaporation. Snow that falls in Russia or Canada has much less $H_2{}^{18}O$ than rain that falls in Malaysia or Peru. Similarly, snow falling at the centre of an ice sheet will have less ^{18}O than snow falling at the edges of the ice sheet, which is due to the preferential condensation of ^{18}O and $H_2{}^{18}O$ precipitating first. From the $\delta^{18}O$ ratio we can infer the temperature of precipitation, and hence how much warmer or colder the Earth was at the time the snow fell.

Additionally, study of Antarctic ice cores examining the ratio of oxygen to nitrogen in bubbles within the ice can be used to infer the level of insolation (solar radiation intensity). These bubbles within the ice can also be analysed to determine the concentrations of greenhouse gases at the time, such as carbon dioxide (CO_2) and methane (CH_4). Resultant changes in climate from differing greenhouse gas concentrations alter the patterns of global evaporation and precipitation, and therefore change the $\delta^{18}O$ ratio. Data from Vostok Station in Antarctica (shown in figure 1.7) shows that the Earth's climate has varied considerably over previous millennia, with peaks and troughs in temperature.

Importantly, we can see the temperature profile closely follows the concentrations of CO_2 and CH_4. Furthermore, the temperature profile follows some periodicity related to the level of insolation. Greater insolation means greater evaporation of water, and hence more water vapour in the atmosphere, which is the most important greenhouse gas. But why does the level of insolation vary with time, and why is there periodicity?

Unlike early pictures of the solar system, the planets do not hold circular orbits around the Sun. The Earth spins on its axis and orbits the Sun; however, there are

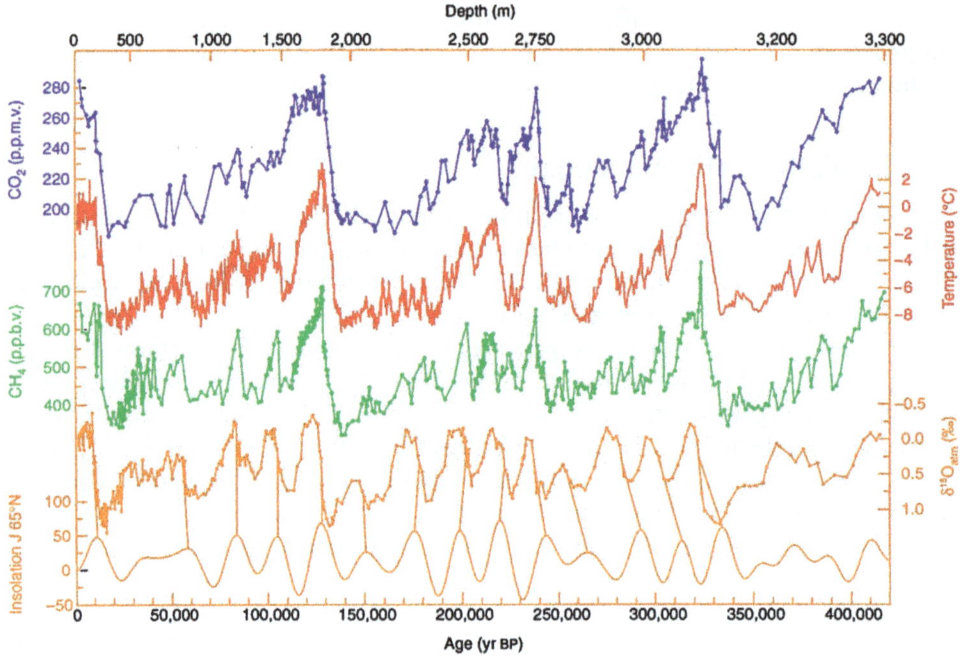

Figure 1.7. Ice core data from Vostok Station in Antarctica, showing $\delta^{18}O$ and associated temperature estimation. Also shown are insolation, CH_4 (parts per billion by volume), and CO_2 (parts per million by volume) concentration from gas bubbles within the ice cores. (This Vostok ice core data image has been obtained by the author (s) from the Wikimedia website [https://commons.wikimedia.org/wiki/File:Vostok_420ky_4curves_insolation.jpg], where it is stated to have been released into the public domain. It is included within this article on that basis.)

several quasi-periodic interactions due to the gravitational forces from other celestial bodies. The Serbian mathematician Milutin Milanković (1879–1958) studied the changes in the Earth's orbital eccentricity, obliquity, and precession, and their influence of the level of insolation. These observed cycles in insolation, and hence cycles in the Earth's temperature, are now generally referred to as Milankovitch Cycles.

Unlike our idealised example earlier in this chapter, the Earth's orbit is not circular, and the distance from the Sun to the Earth is not a constant. The Earth's orbit is an ellipse. Eccentricity is a measure of the departure of an ellipse from being circular (see figure 1.8); the eccentricity of a circle is by definition 0, while a parabola has an eccentricity of 1 and an ellipse has an eccentricity <1. The equation for a circle is of the form

$$x^2 + y^2 = r^2$$

while an ellipse has the form

$$\frac{x^2}{a^2} + \frac{y^2}{b^2} = 1$$

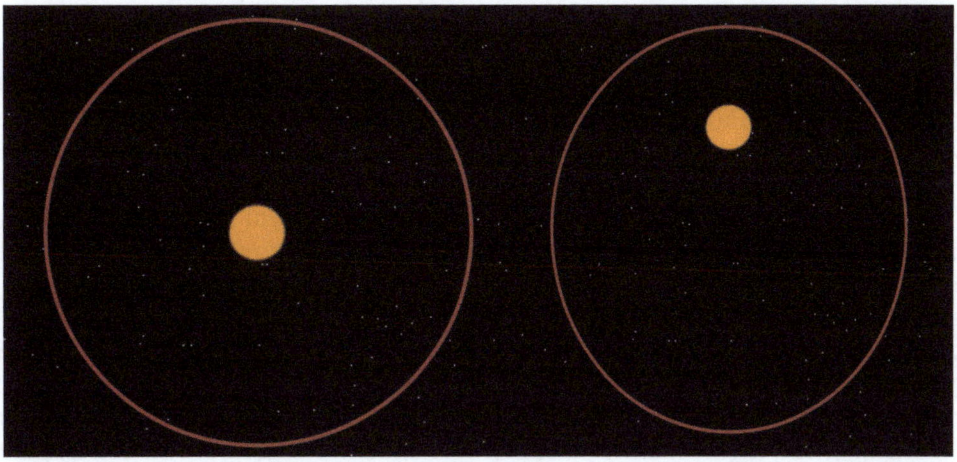

Figure 1.8. Illustration of a circular orbit (left), and an elliptical orbit with highly exaggerated eccentricity (right).

and has an eccentricity (e) of the form

$$\sqrt{1 - \frac{b^2}{a^2}}.$$

The shape of the Earth's orbit varies with time between nearly circular, with a low eccentricity (e = 0.000 055), and a more elliptical orbit (e = 0.0679). The mean eccentricity of the Earth's orbit is considered to be e = 0.0019. Figure 1.8 provides an illustration of both a circular and a highly eccentric elliptical orbit.

If the Earth was the only planet orbiting the Sun, then the eccentricity of the orbit would not perceptibly alter over millions of years. However, the presence of other large planets in our solar system such as Saturn and Jupiter produces cyclic interactions due their gravitational field, which vary with their orbit around the Sun and the Earth's position relative to them. As you might expect with seven other planets and dwarf planets (including: Pluto, Ceres, and Eris), predicting variations in eccentricity due to so many gravitational interactions is complex. However, the signal is dominated by the largest bodies, resulting in a few main components for the periodicity of the eccentricity. The main component has a period of 413 000 years ($e \pm 0.012$). While other components have periods of 95 000 years and 125 000 years, resulting in a beat frequency of ~400 000 years. The combinations of these signals loosely combine into a 96 000-year cycle (see figure 1.12) with a magnitude between −0.03 to +0.02. Presently the Earth's orbit has an eccentricity of e = 0.017 and is decreasing.

As eccentricity increases, variations in seasons increase due to varying distance from the Sun and varying insolation. However, this variation in insolation due to eccentricity is generally small, the intensity of the seasons is not primarily governed by the distance from the Sun, but rather the axial tilt of the Earth. The relative increase in insolation at the perihelion (the closest approach to the Sun) compared to insolation at the aphelion (furthest distance from the Sun) is roughly 4× the eccentricity. Currently this amounts to a variation in insolation of ~6.8%. Perihelion currently occurs around the third of January, while aphelion is around the fourth of July.

Changes to the Earth's eccentricity do not change the length of a year or alter Earth's motion along its orbit. However, higher eccentricity does exaggerate behaviour due to precession and axial tilt, which are explained below.

As we all know, the Earth's axis of rotation is not upright (perpendicular to the direction of orbital motion); instead, the Earth spins around an axis ~23° off perpendicular (figure 1.9). It is this axial tilt (obliquity) that is responsible for the seasons we experience and the long polar nights and days. The angle of the Earth's axial tilt varies over time with respect to the planet's orbital plane. Axial tilt varies by up to 2.4° between 22.1° and 24.5° in a periodic way, taking approximately 41 000 years. When the obliquity increases, the magnitude of the seasonal variations in insolation increase. Summers receive more radiative flux and winters comparatively less. The opposite is true as obliquity decreases. It should be noted that these changes to the magnitude of insolation are not equal over the entire Earth's surface. At higher latitudes, the mean annual insolation increases with obliquity, while at lower latitudes obliquity decreases insolation. As such, it can be argued that lower obliquity favours the onset of ice ages by reducing both the overall mean summer insolation and additionally reducing insolation at higher latitudes. This results in reduced melting of the previous winter's precipitation, as most of the Earth's snow

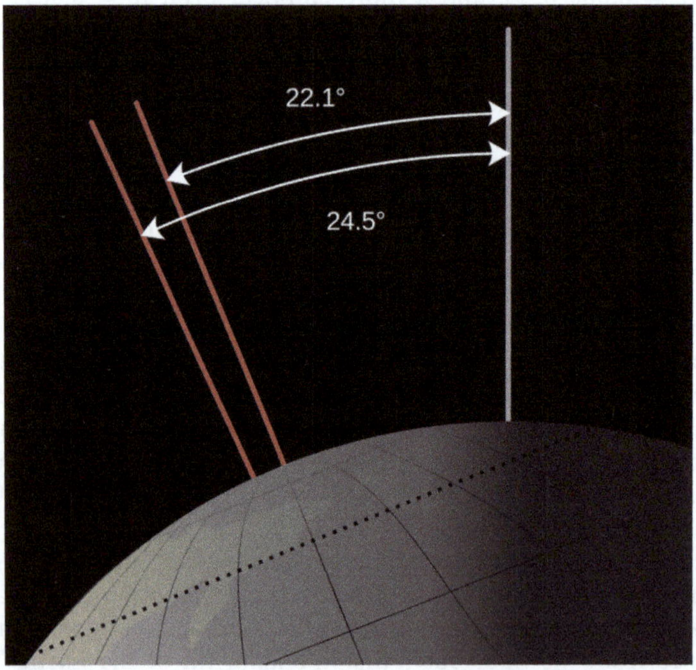

Figure 1.9. Illustration of changes to the Earth's axial tilt or obliquity (orbital motion is out of the page) (courtesy of NASA).

and ice lies at higher latitudes. The resulting change in the Earth's albedo can result in further cooling, creating an ice age, which will persist until the obliquity increases.

Currently, the obliquity of the Earth is 23.44°, about halfway through its cycle, and is decreasing. The obliquity will reach its minimum around 11 800 AD, and was last at its maximum in 8700 BC. Therefore, we can say that the Earth is currently experiencing a cooling trend.

The Earth not only spins about a tilted axis, which varies with time but also 'wobbles' like a spinning top. This process termed precession is the trend in the direction of the Earth's axis of rotation to rotate relative to the orbital plane and the fixed stars (see figure 1.10). This axial precession has a period of roughly 26 000 years, and is due to gravitational tidal forces exerted by the Sun and the Moon upon the Earth. The contribution from each is roughly equal.

When the Earth's axis of rotation points towards the Sun at perihelion (i.e. the North Pole is pointed toward the Sun), the Northern Hemisphere has a greater variation in seasons while the Southern Hemisphere experiences reduced seasonal variability (milder seasons). Meanwhile, the inverse is true when the axis of rotation points away from the Sun at perihelion (i.e. South Pole points towards the Sun),

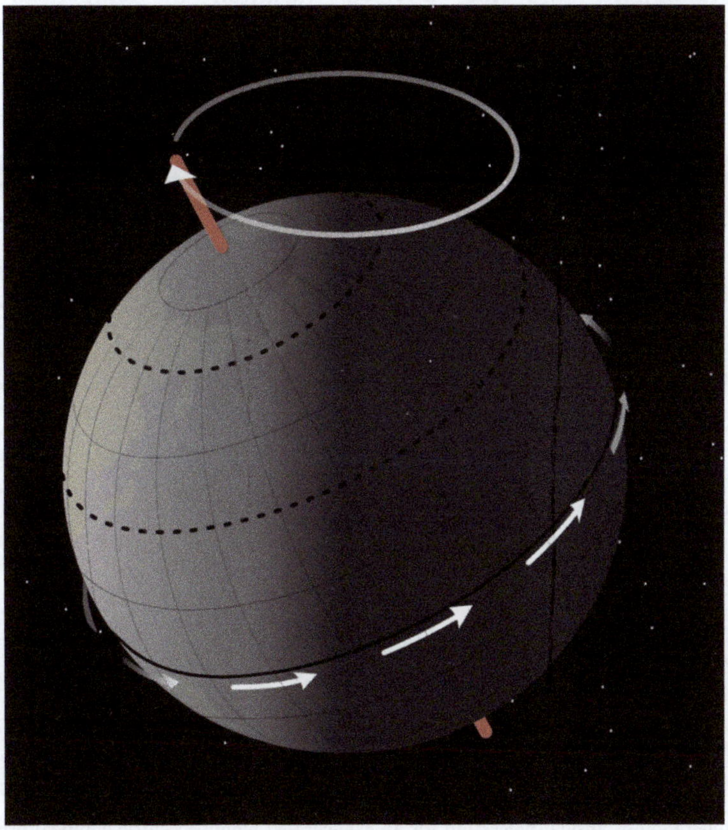

Figure 1.10. Illustration of the Earth's axial precession (courtesy of NASA).

resulting in greater seasonal variations in the Southern Hemisphere and milder seasons in the Northern Hemisphere. The hemisphere that is at perihelion in summer will receive most of the increase in insolation, resulting in higher summertime temperatures; however, this hemisphere will experience much colder winters when it is in aphelion. this present, the Southern Hemisphere is in perihelion during the summer and aphelion during the winter, and thus experiences a greater variation in seasonal extremes relative to the Northern Hemisphere, which experiences milder seasons.

In addition to eccentricity, obliquity, and axial precession, the Earth's elliptical orbit precesses in space (see figure 1.11). This orbital (apsidal) precession completes a cycle every 112 000 years relative to the fixed stars. Apsidal precession occurs in the elliptic plane, and alters the Earth's orbit relative to the elliptic. In combination with changes in eccentricity, this alters the length of seasons.

All of these changes in the movement of the Earth relative to the Sun alter the amount of insolation the Earth receives. The sum of all these different cycles in insolation produces a complex signal of temperature oscillation, which can be observed in both the fossil record and in ice cores. Figure 1.12 shows the cyclic contributions of the different orbital movements (Milankovitch cycles), the resultant changes in insolation and the observed variation in $\delta^{18}O$ from ice cores and within benthic foraminifera and the inferred variation in temperature. Benthic Foraminifera (forams for short) are single-celled organisms related to amoeba, and produce a hard shell. Foram shells are made of calcium carbonate ($CaCO_3$) and are found in many common geological environments. The $\delta^{18}O$ ratio in the shell is

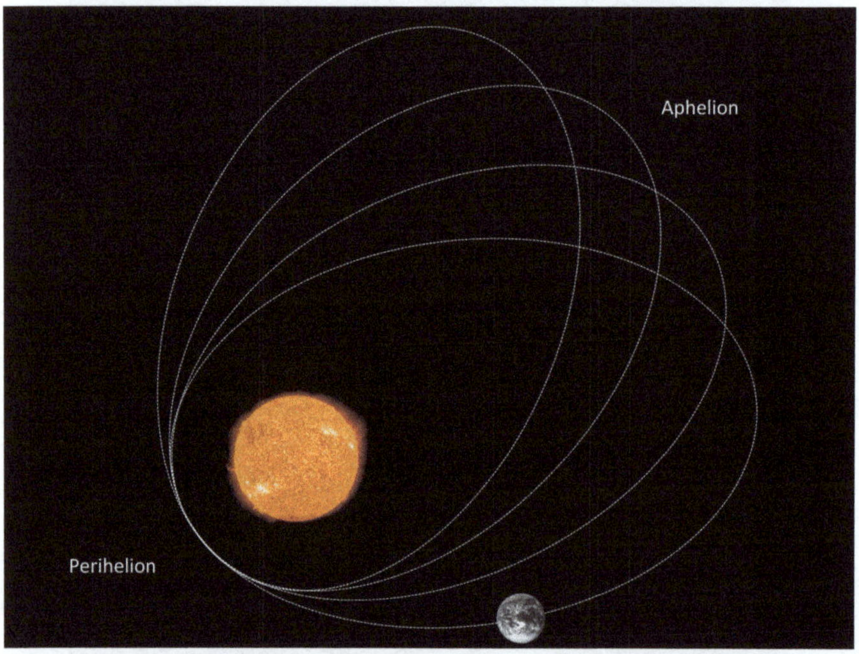

Figure 1.11. Illustration of apsidal precession for an exaggerated highly eccentric elliptical orbit. Animation available from https://doi.org/10.1088/978-0-7503-1197-7.

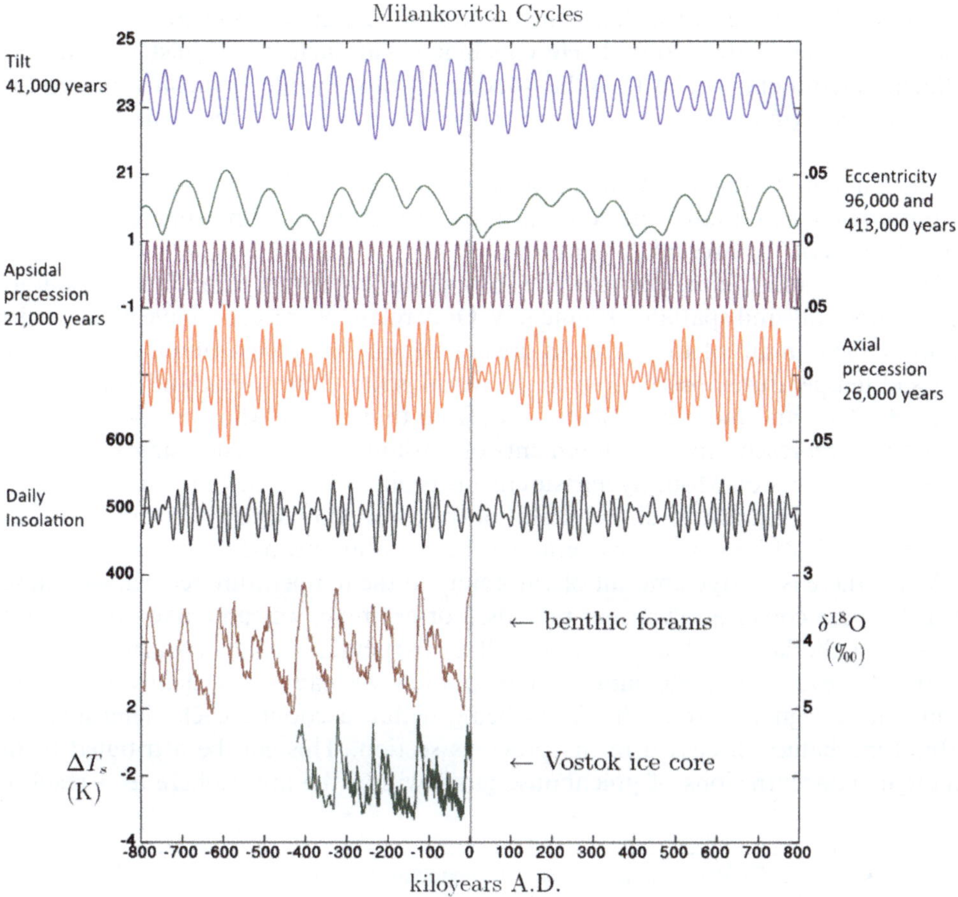

Figure 1.12. Plot of the various components of Milankovitch cycles and the observed $\delta^{18}O$ and temperature from ice cores and the fossil record. This Milankovitch Cycles image has been obtained by the author from the Wikimedia website where it was made available under a CC BY-SA 3.0 licence. It is included within this article on that basis. It is attributed to Incredio.

used to determine the temperature of the surrounding water at the time the organism was alive. This can be used as another indicator along with ice core $\delta^{18}O$ and gas bubbles within the ice to determine what conditions were like on Earth in the past.

1.3 The anthropogenic greenhouse effect

As we have explored in the previous section, the makeup of the Earth's atmosphere is responsible for the climatic conditions we experience and for life as we know it. Given the vast volume of the atmosphere it seems unlikely that human activities could influence the composition of the atmosphere and alter the energetic balance of the planet. The natural greenhouse effect was first described in 1859 by the British scientist John Tyndall, when he discovered that the most common components of the atmosphere—nitrogen and oxygen—were transparent to both visible and infrared radiation. Whereas gases such as carbon dioxide, methane, and water vapour were not transparent in the infrared. He concluded that such gases must have a great influence on our climate. In 1894 the Swedish chemist Svante Arrhenius showed that anthropogenic (man-made) emissions had the potential to alter the climate by further reducing the transparency of the atmosphere in the infrared spectrum. He further concluded that at the current rate of emissions that it would take mankind 3000 years of burning coal to double the concentration of CO_2 in the atmosphere; in this last point he was off by around 28 centuries!

While we can examine past climate through ice cores and forams, the resolution both temporally and spatially is quite low. Only relatively recently have we started to monitor the climate of the Earth in greater detail. The oldest running continuous series of temperature observations in the world is the Central England Temperature Record. Daily and monthly temperatures from three observations stations are used to produce representative measurements of a triangular area enclosing Lancashire, London, and Bristol. Monthly measurements begin in 1659, and daily measurements begin in 1772. Figure 1.13 shows a plot of the mean annual temperature from 1659 to the end of 2015, created by averaging the monthly means.

While there is a large amount of variability in the temperature record, we can see that there are cooler and warmer periods. For example, we can make out the 'mini ice-age' of the later 17th century, as well as particularly cold or warm individual years. However, since the industrial revolution we can see a steady rise in the temperature signal, despite the Earth being within a cooling cycle determined by orbital mechanics discussed in the previous section. This can be attributed to the changing concentrations of greenhouse gases within the atmosphere as a result of

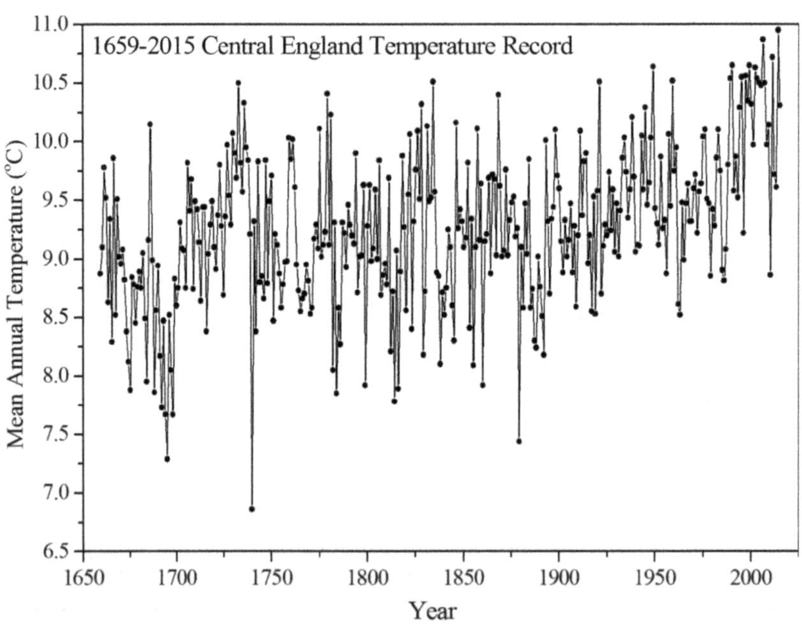

Figure 1.13. Plot of mean annual temperature from the Central England Temperature record.

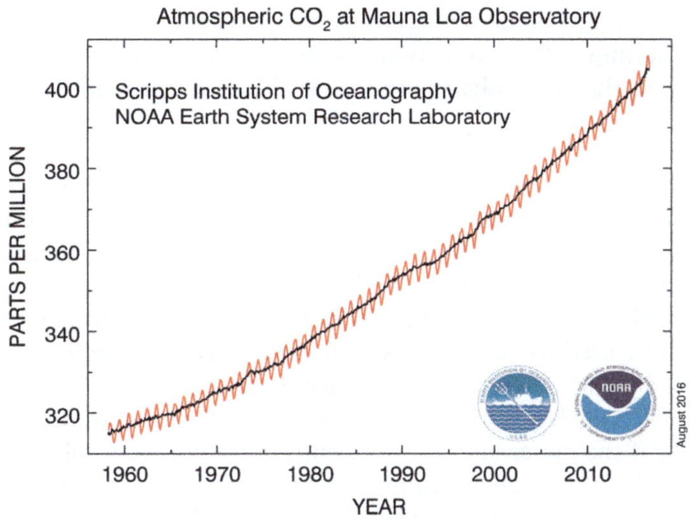

Figure 1.14. Plot of monthly measurements of CO_2 from Mauna Loa observatory, the red line represents monthly measurements while the black line is an annual mean (courtesy of NOAA).

human activities, enhancing the already present natural greenhouse effect. Figure 1.14 shows the trend in atmospheric CO_2 since 1958, the so-called Keeling curve, based on the work started by Charles Keeling. Monthly measurements of atmospheric CO_2 concentration at the Mauna Loa observatory (Hawaii) by the National Oceanic and Atmospheric Administration (NOAA) show an accelerating trend in CO_2 concentration. Recently, atmospheric concentration passed 400 ppm (parts per million) for the first time since modern humans have walked the Earth (compare figure 1.14 to the concentrations shown in figure 1.7 from ice cores).

The oscillations in the atmospheric CO_2 (red line) are a result of the seasonal variations in the northern hemisphere. Since the majority of the Earth's land mass and forests are located in the Northern Hemisphere, the atmospheric CO_2 concentration is dominated by the northern summer and winter, due to the annual cycles of plant life. Worryingly, this trend in CO_2 concentration does not show any visible deviations due to the Rio Earth summit (1992), the Kyoto protocol (1997), or Rio +20 (2012); the only step in the data is attributed to the collapse of the Soviet Union in the early 90s.

There is often much confusion about the origin of this additional CO_2: whether it comes from volcanoes, deforestation, or from the burning of fossil fuels. There is evidence however, to show that the carbon emissions increasing the CO_2 concentration of the atmosphere are a result of anthropogenic emissions, primarily from the burning of fossil fuel. Like oxygen, carbon exhibits several different isotopes, with different masses. Carbon in the atmosphere is ~99.89% carbon-12 (^{12}C) and ~1.11% carbon-13 (^{13}C), and trace amounts of carbon-14 (^{14}C). ^{12}C and ^{13}C are stable, but ^{14}C is radioactive with a half-life of 5730 years (a half-life is the amount of time it takes for the radioactivity of a substance to halve as it decays). The length of this half-life means that any ^{14}C that was created when the Earth formed would long since have disappeared, implying that new ^{14}C must be constantly being made. This process occurs through highly energetic particles, called cosmic rays, entering the atmosphere, resulting in the production of free neutrons. These neutrons collide with nitrogen atoms in the atmosphere, producing ^{14}C and a proton.

$$^{1}n + {}^{14}N \rightarrow {}^{14}C + {}^{1}p.$$

This process occurs in the troposphere and stratosphere at an altitude of 9–15 km. The ^{14}C combines with oxygen molecules to from radioactive $^{14}CO_2$ at a concentration of one part in a trillion (1:10^{12}), which is taken up by plants and algae, passing up the food chain to other living organisms. While the organism is living, this concentration is maintained, only reducing once the organism dies. This means that the amount of ^{14}C present can be used to date objects such as wooden axe handles or animal hides used for shelter by early humans, or any carbonaceous material up to about 60 000 years old. By carbon dating the atmosphere, we can determine the source of the additional CO_2. If the increasing CO_2 concentration were the result of deforestation, then the amount of ^{14}C would remain constant, whereas we observe a decreasing concentration of ^{14}C in the atmosphere. This shows that the carbon emissions are coming from an old source of carbon, such as the burning of fossil fuels, which are old enough that all ^{14}C in the material has long since decayed. We can track the concentration of ^{14}C in the atmosphere over recent times through observation of tree rings, and find that the decreasing concentration of ^{14}C coincides with the industrial revolution and has decreased in-line with increasing fossil fuel use. While volcanoes do produce substantial amounts of CO_2, and much of this would be old carbon with reduced ^{14}C, these events are not regular enough to explain the observed trends. Additionally, the eruption of volcanoes needs to be weighed against the disruption they cause to normal human activities. The recent eruption of the Icelandic volcano Eyjafjallajökull in 2010, could be considered a carbon negative event due to the widespread disruption to air travel across Europe. Hence, we can conclude that the observed trends in atmospheric CO_2 concentration are primarily a result of human activities. An additional source of confusion when considering the anthropogenic greenhouse effect comes from the inertia of the carbon cycle and the atmosphere. It takes a long time for changes to the atmosphere to exhibit themselves as warming at the surface of the planet because both the land and oceans have high thermal inertia. We are only just starting to observe oceanic warming despite having been pumping out increasing amounts of CO_2 for over two centuries. Figure 1.15 overlays the observed CO_2 concentrations shown in figure 1.14 on top of the observed global temperature anomaly (variation). The baseline (0 °C) for this temperature anomaly is the 20th Century (1901–2000). Through close examination of the trends, we can see the lag between increasing CO_2 concentrations and increasing temperatures.

This thermal inertia also has a downside that even if we were to stop all carbon emissions tomorrow, the land surface and oceans would continue to warm and sea levels would still continue to rise, until equilibrium is reached (see section 1.1). Of course, CO_2 is not the only greenhouse gas; water vapour is the most important in terms of the elevated temperatures the Earth experiences, but water vapour is not directly produced by human activities. There are, however, several other gases that have a significant greenhouse effect, notably methane, which is produced by

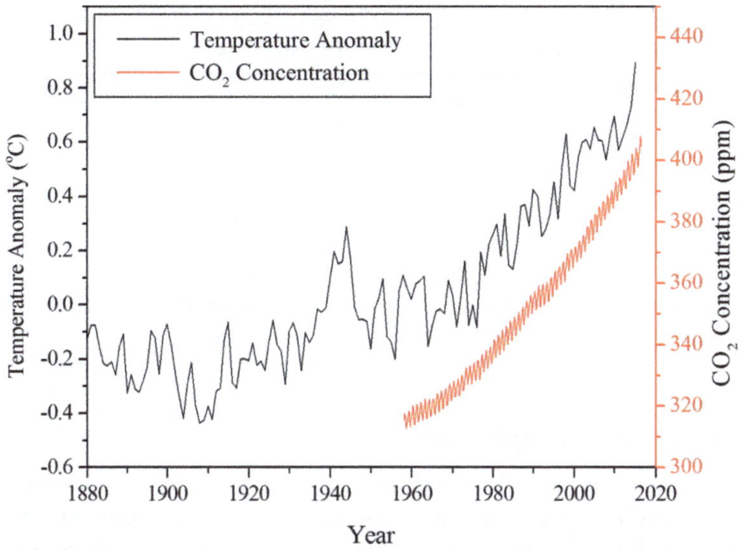

Figure 1.15. Plot of average land and ocean temperature anomaly and CO_2 concentration against time (data obtained courtesy of NOAA).

extensive farming, particularly of cattle or rice. There are also several man-made chemicals that are very effective at trapping heat in the atmosphere, such as chlorofluorocarbons (CFCs), sulphurhexafluoride (SF_6), and hydrofluorocarbons (HFCs). Not only do these different chemicals have different effectiveness at absorbing heat radiated from the Earth's surface, but they remain in the atmosphere for different amounts of time. A CO_2 molecule typically remains in the atmosphere for between 5 and 200 years (generally accepted as 100 years) before it is absorbed by a plant and starts to move around the carbon cycle. Methane is only present in the atmosphere for around 12 years, but it is much more effective at trapping heat. Thus, we need a measure to compare the relative effectiveness of these different chemicals in order to inform mitigation policies. Relative global warming potential (GWP) is used to make this comparison; CO_2 is used as a baseline and has a GWP = 1 for a 100 year horizon. Table 1.1 compares the different GWP of several important anthropogenic greenhouse gases.

This also allows the equation of different greenhouse gas emissions as carbon dioxide equivalent (CO_2e or CO_2-eq). Increasing the atmospheric concentration of any of these gases will have a warming effect as more heat radiated from the Earth's surface will be intercepted and re-radiated back at the surface. The term radiative forcing is used to quantify the additional amount of warming experienced. A doubling of CO_2 above pre-industrial levels (278 ppm) is expected to have a radiative forcing of $\sim$4 W m^{-2}. Comparing this to the values shown in figure 1.6, we can see that this will have a profound effect on the climate we experience.

Table 1.1. Lifetime and relative GWP of different atmospheric chemicals against a 100 year baseline of CO_2 accounting for climate-carbon feedbacks. (Source IPCC AR5 and AR4 for SF_6).

Name	Formula	Lifetime	Relative GWP
Carbon dioxide	CO_2	100	1
Methane	CH_4	12.4	34
Nitrous oxide	N_2O	121	298
Hydrofluorocarbon	HFC-134a	13.4	1550
Chlorfluorocarbon	CFC-11	45	5380
Carbon tetrafluoride	CF_4	50 000	7350
Sulphur hexafluoride	SF_6	3200	22 800

1.4 Climate change projections

The Intergovernmental Panel on Climate Change (IPCC) assessment reports summarise global climate change impacts. We will discuss projections of climatic change in greater detail in the following chapters of this book. Here we will briefly summarise some of the projections from the IPCC, and their origins.

For the first four IPCC assessment reports (1990–2007), estimates of future climate change have been based upon socio-economic scenarios, detailed in the IPCC 'Special Report: Emission Scenarios' (SRES). Future greenhouse gas emissions are a product of the complex interactions of many different dynamic systems. These SRES scenarios cover a wide range of driving forces that influence current and future emissions, including demographic, technological, and economic developments. These scenarios include the range of emissions for all the relevant greenhouse gases and their driving forces. As such, the IPCC state that the likelihood of any single emissions path actually occurring is very small. Therefore, when dealing with future climate change, we are considering possible projections of what the world could be like, rather than definitive predictions of what the world will in 20, 40, or 100 years' time.

By the end of this century, the world will have changed in ways that are very difficult to imagine, in the same way those who lived at the turn of the last century would find it hard to predict today's way of life. The SRES scenarios consider four different storylines to describe consistently the interrelationships between emission driving forces and their evolution. Each storyline (A1, A2, B1, and B2) represents different demographic, social, economic, technological, and environmental developments. These storylines become increasingly divergent, in irreversible ways. Furthermore, because of their complexity, their feasibility cannot simply be based upon extrapolation of current socioeconomic trends.

- The A1 storyline describes a world of rapid economic and population growth, and rapid introduction of new and more efficient technology. Global population peaks in the middle of the century and declines thereafter. The major underlying themes are increased social and cultural interactions, and convergence between regions with reduction in the differences in regional per capita income. The A1 family is split into three distinct scenarios: fossil fuel intensive (A1FI); non-fossil energy sources (more technological) (A1T); or a balance across all energy sources (A1B).

- The A2 storyline describes a heterogeneous world where self-reliance and preservation of local identity is key. Population growth rates across all regions converge very slowly, leading to continuously increasing population. Technological advancement is slower and more fragmented than in other storylines, and economic development is primarily associated with regional and per capita economic growth.
- The B1 storyline describes a convergent world with a population that peaks in the middle of the century and declines thereafter (as in A1). There are rapid changes in economic structures towards a service and information economy, with associated reductions in material intensity and the introduction of clean resource efficient technologies. The emphasis is upon global solutions to economic, social, and environmental sustainability and equity, without additional climate initiatives.
- The B2 storyline describes a world where the emphasis is on local solutions to economic, social, and environmental sustainability. The global population continuously increases, but at a slower rate than A2, with an intermediate level of economic development. Technological advancement is more diverse than in the B1 and A1 storylines. The scenario is orientated towards environmental protection and social equity, but at a more regional local level.

No probability or likelihood is associated with any of these storylines or scenarios. Some storylines, such as B2, are becoming increasingly infeasible due to recent population growth, but other factors within the storyline are still possible. Hence, this storyline cannot simply be ignored. The primary issue with the SRES storylines is their age. Created in 1990, they do not account for the rise of countries such as China and India as massive economic powers and the associated demographic, social, and local technological improvements implemented, or the resultant carbon emissions. For this reason, a new set of scenarios were created by the scientific community and implemented in the most recent IPCC 5th Assessment report, the so-called 'Representative Concentration Pathways' (RCPs).

There are four RCPs, each covering the period 1850–2100 with extensions (extended concentration pathways (ECPs)) formulated for up to 2300. The RCPs are named according to the radiative forcing level (in W m^{-2}) at 2100, i.e. RCP 2.6, RCP 4.5, RCP 6, and RCP 8.5. The RCPs represent a simpler set of scenarios compared to SRES: instead of four main storylines, each is augmented with a range of possible scenarios based upon different socioeconomic futures. The RCPs are simply related to a level and acceleration of radiative forcing to avoid ambiguity. The science behind them is actually more complex than was used for the SRES storylines. The four RCPs are consistent with certain socio-economic assumptions, and are anticipated to provide flexible descriptions of different possible social, economic, demographic, and technological futures. Each RCP has been shown to be achievable via several different regional and global socio-economic routes. In this way, these new RCP scenarios are devolved from socio-economic factors, to help avoid confusion and dismissal by the general public. For similar reasons, four pathways were chosen rather than three to avoid the perception that the middle option is best and the safest bet. Table 1.2 provides a summary of the key features of the different RCP scenarios.

The details and relevance of the different social, economic, demographic, and technological aspects of can be hard to comprehend. Essentially, all of these aspects can be reduced down to anthropogenic CO_2-eq emissions and a level of radiative forcing. Figure 1.14 shows changes to atmospheric CO_2 concentration form 1958 to the present day, while figures 1.16 and 1.17 extend this trend in atmospheric CO_2 for the different SRES and RCP scenarios up to the end of the century.

We can see that despite the new science and inclusion of more recent socio-economic data, such as the rise of China and India as major economic powers and the associated emissions, there is little difference between the higher end scenarios (A1FI and RCP 8.5) by the end of the century. For brevity, the following chapters of this book will use climate change projections based upon RCP 8.5, except in the case of more regional projections, which have not yet been updated, where A1FI will be used.

The UK has among the most detailed climate change projections in the world. The current iteration of these projections, compiled by the Met Office Hadley Centre, were released in 2009 and are termed UKCP09 (UK Climate Projections 2009). These predate the release of the IPCC 5th Assessment report, and hence are based upon the SRES emissions scenarios. However, they differ from many climate change projections in that they are probabilistic.

The UK Climate Projections 2009 are not solely based upon simulations of the UK Met Office Hadley Centre climate models; instead, they are based upon many different climate models from around the world. This collection of climate models is run many times with different estimates for uncertain processes in the models (referred to as perturbations). Uncertainty in climate models arises from the natural variability in the climate system, which is inherently random, our incomplete understanding of all of the Earth's climate processes, as well as uncertainty in future greenhouse gas emissions. This allows the collection of climate models to create a spread of results based upon the limitations of our current understanding. By combining the results into distributions of possible values or *probability density functions (PDF)* we can provide information about the relative likelihood of different

Table 1.2. Summary table of the representative concentration pathways.

	Greenhouse gas emissions	Agricultural area	Air pollution
RCP 2.6	Very low	Medium for cropland and pasture	Medium-low
RCP 4.5	Medium-low, very low baseline	Very low for cropland and pasture	Medium
RCP 6	High mitigation, medium baseline	Medium for cropland, very low for pasture (total low)	Medium
RCP 8.5	High baseline	Medium for both cropland and pasture	Medium-high

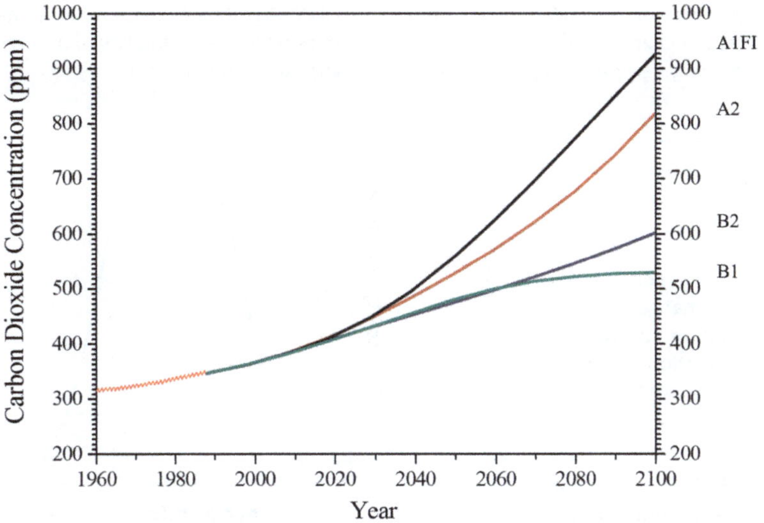

Figure 1.16. Global CO_2 concentration from 1960–2100 for four emissions scenarios. The range of uncertainty for each scenario is not shown.[1]

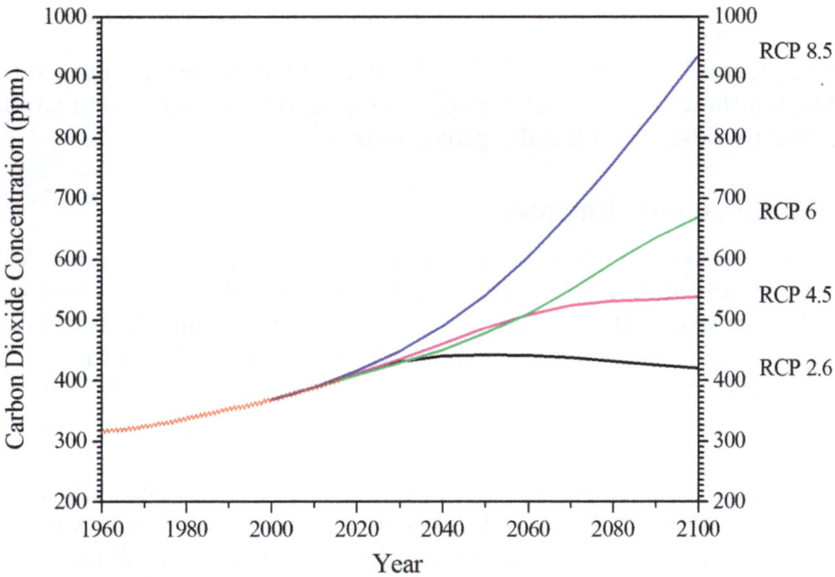

Figure 1.17. Global CO_2 concentration from 1960–2100 for the four RCPs, overlaid with Mauna Loa historic CO_2 data.

future outcomes. This process is demonstrated graphically in figure 1.18 by comparing UKCP09 to the previous iteration of UK climate projections UKCIP02.

The next iteration of the UK climate projections is planned to be released in 2018 (UKCP18) and will be based upon the RCP scenarios. Full details have not

[1] Image adapted from [1].

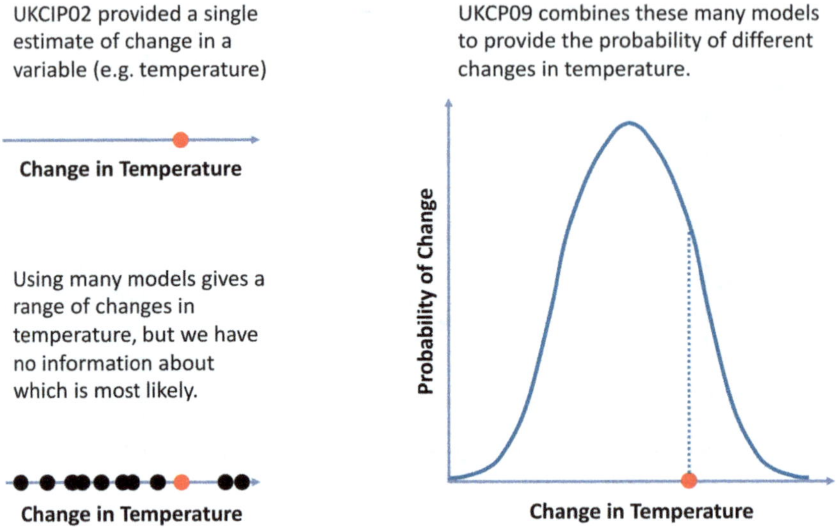

Figure 1.18. Illustration of how many climate model outputs are combined to form probabilistic estimate of climate change in UKCP09[2].

yet been made public but UKCP18 will consist of probabilistic data similar to UKCP09, but there will be greater emphasis on the representation and handling of extreme weather events within the projections.

1.5 Climate change impacts

Within this book we talk about climate change and not global warming. Although the two terms are often used interchangeably in the media, this is not necessarily correct. The term global warming implies a global warming effect and could be interpreted to simply mean warmer temperatures, which, to those of us who live at higher latitudes, might be viewed as a good thing. However, the climate system is complex, and the weather we experience is driven by atmospheric convection currents, the magnitude of which is a function of the rotation of the Earth and the difference in temperature between the surface and the upper atmosphere and between the equator and the poles. As some areas warm, other areas are expected to cool, the difference in temperature between the land and oceans will increase due to different thermal inertias altering the weather experienced in different global locations. Changes to these can disrupt and alter these convection currents, changing global weather patterns. Therefore, the term global warming is not strictly true. Conversely, climate change is not just temperatures, but also sea level rise, changes to storms and monsoons, the persistence of droughts, or the melting of permafrost to provide arable land. Figure 1.19 shows an illustration of the atmospheric convection currents, which drive global weather patterns; each

[2] Image adapted from [2].

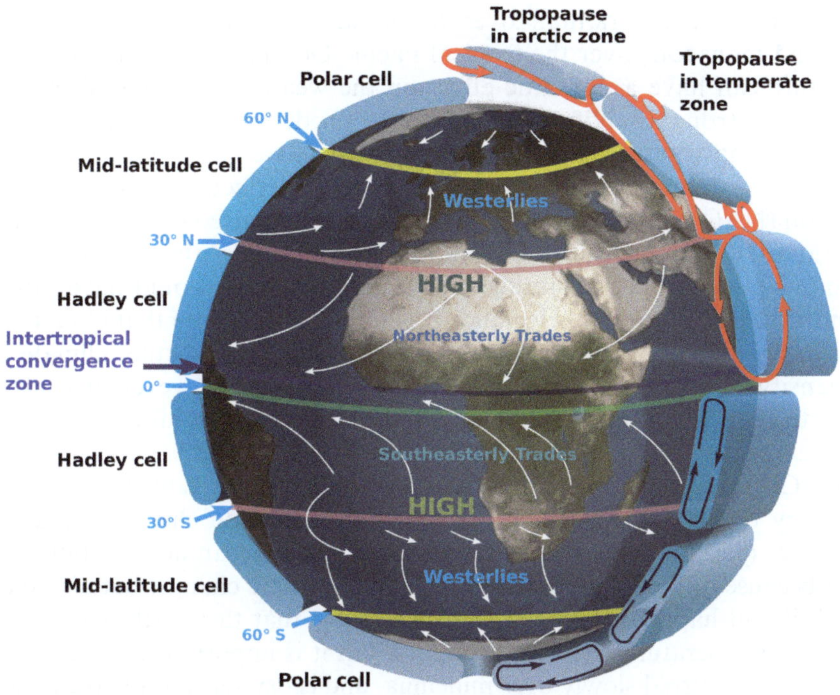

Figure 1.19. Illustration of the atmospheric convection currents that drive our climate system (courtesy of NASA).

hemisphere has three convection cells, the Polar cell, the Mid-latitude or Ferrel cell, and the Hadley cell.

There is limited mixing between the atmospheric convection cells, and air exchange might only occur during certain conditions. For example, the ozone hole in the Antarctic polar cell is a result of limited exchange between the southern Polar cells and the Ferrel cell during the cold Antarctic winter, the so-called Polar Vortex. Furthermore, we can see from figure 1.19 that at certain places air is rising or falling, not travelling horizontally. It is for this reason that it was difficult for sailboats to cross the equator, as air rises vertically here and there is little horizontal wind giving rise to the 'doldrums'. Likewise, at the horse latitudes around 30° between the Ferrel cell and the Hadley cell, air is falling. These atmospheric convection currents are also responsible for the trade winds at sea and the prevailing winds we experience on land.

There is a similar pattern at sea. The global ocean convection currents, which carry warm water near the surface from the tropics towards the poles where it cools and sinks, are responsible for the timing, magnitude, and trajectory of monsoons in India, hurricanes in the Caribbean, and storms in the UK. The temperature of the ocean surface determines all these things. The El Niño Southern Oscillation is the warming of the ocean surface in the central and eastern tropical Pacific Ocean (La Niña refers to the cooling trend). This warming shifts the atmospheric circulation slightly, thus reducing rainfall over Indonesia and Australia while increasing rainfall and monsoons over the tropical Pacific Ocean. Changes to ocean surface temperature can have a dramatic effect on the weather experienced by different countries. A strong El Niño event recently produced a prolonged drought in Australia lasting several years, threatening Australian wine production along the Darwin. The Atlantic Ocean also exhibits a similar effect, the Northern Atlantic Oscillation (NAO). The series of intense storms experienced during the summer of 2012 in the UK were a result of a warmer than normal Atlantic surface temperature moving storm tracks southwards, which otherwise would have missed the UK and travelled on towards Iceland. It is important to note that climate change science has so far been unable to reach a consensus as to whether climate change will intensify or reduce the El Niño and NAO effects. However, it is clear that changes to the status quo has the potential to have very profound influences on global weather.

The IPCC states that 'the resilience of many ecosystems is likely to be exceeded this century by an unprecedented combination of climate change and associated disturbances (e.g. flooding, drought, wildfire, insects, ocean acidification)'. This is not just because of the magnitude of anticipated climate change, but also the speed at which it will happen. We have seen previously that the Earth has experienced variation in temperature and climate before, but it is important to understand that these changes occurred slowly over millennia, and ecosystems were able to adapt or migrate in response. The IPCC provide detailed estimates of the global impacts of a changing climate, some of which are summarised below.

Increased CO_2 in the atmosphere and warmer temperatures will lead to increased plant growth, and hence increased CO_2 uptake. However, if temperatures rise too high then plants begin to respire just like every other creature. Over the course of this century, net carbon uptake by terrestrial systems is expected to peak before the middle of the century, and then weaken or even reverse, leading to amplification of climate change. It is estimated that 20%–30% of plant and animal species are likely to be at risk of extinction if the average global temperature increase is greater than 1.5 °C–2.5 °C. For temperature increases exceeding this range, there are expected to be major changes to ecosystems and the interactions between species and a migration of species. This will result in a reduction in biodiversity, and will have impacts on water and food supplies.

Increased plant growth will lead to increased crop productivity at mid to high latitudes, with increased productivity depending upon temperature increase (1 °C–3 °C) and the crop in question. However, this will be balanced by reduction in crop productivity at lower latitudes. In seasonally dry and tropical regions, crop productivity is expected to decrease for even small increases in temperature (1 °C–2 °C). Globally, food production is projected to increase for a 1 °C–3 °C temperature increase, but decrease thereafter.

Climate change will also have the potential to affect the health of millions of people. Reductions in crop productivity would increase malnutrition and increase the burden of diarrhoeal diseases. Changes to weather patterns resulting in more severe storms and extreme weather events increase risk of physical injury. Changes to weather and climate will alter the global distribution of infectious diseases. In urban areas, increased concentrations of ground level ozone increases risk of cardio-respiratory diseases. There will be benefits as well; in temperate zones such as the UK, cold-related deaths in winter currently far outweigh heat-related illness in summer. Climate change will reduce the number of cold-related deaths in these areas, but higher summertime temperatures and increased instances and intensities of heat waves will result in more deaths in summer. In the short term, temperate areas should see reduction in temperature-related deaths. Globally, however, rising temperatures will result in increased mortality, especially in developing countries.

Changes to the frequency and intensity of precipitation will lead to increased stress on water resources, exacerbated by population growth and land-use change such as urbanisation. Changes to precipitation and temperature will increase surface water run-off. Run-off is projected to increase by up to 40% at higher latitudes and in some wet-tropical areas (e.g. Southeast Asia), but at the same time decrease by up to 30% over dry areas at the mid-latitudes and in dry-tropical areas, due to decreased rainfall and increased uptake and evapotranspiration by plants. Areas that are currently affected by drought are expected to grow in size, having an adverse impact upon agriculture, water supply, and human health. Climate research suggests that there will be a significant increase in future heavy rainfall events for many regions, even regions where the mean total annual rainfall is projected to decrease. This poses challenges to society in the form of floods, damage to infrastructure, and water quality.

Sea level rise due to thermal expansion of the oceans as they warm will threaten many low-lying coastal areas. The greatest increases in sea level will be in the densely populated mega-deltas of Asia and Africa, while small islands are particularly vulnerable and in danger of disappearing completely (e.g. Micronesia). In addition, thermal expansion increases sea levels due to stronger storms and greater storm surges. In the presence of a low-pressure system, the atmosphere exerts less force on the ocean surface, and the level rises. Storm surges in the UK can currently exceed 1 m in height, and stronger storms will be typified by even lower pressure systems and hence greater storm surges. This will pose particular problems for low-lying areas, especially those with dense urban populations. The most vulnerable societies, industries, and settlements are those in coastal and river flood plains, especially those where the economy is dependent on climate-sensitive resources and is sensitive to extreme weather events (e.g. harbours). Rapid urbanisation in these areas will exacerbate the risk from climate change.

The further chapters of this book deal with specific climate change risks and their impact on urban areas. As might be expected, there will be overlap and common solutions to some of the climate change-related issues. The final chapter aims to summarise these and tease out common solutions to increase the resilience of urban areas to climate change.

References

[1] Hulme M, Jenkins G J and Lu X *et al* 2002 Climate Change Scenarios for the United Kingdom: The UKCIP02 Scientific Report, Tyndall Centre for Climate Change Research, School of Environmental Sciences (Norwich, UK: University of East Anglia) ISBN 0 902170 60 0

[2] Jenkins G J, Murphy J M, Sexton D S, Lowe J A, Jones P and Kilsby C G 2009 *UK Climate Projections: Briefing report* (Exeter, UK: Met Office Hadley Centre) ISBN 978 1 906360 04 7

Chapter 2

Water

Over recent years, people in the UK have experienced increasing weather variability, stronger storms, and more intense rainfall. There is perhaps still too little evidence to lay these events solely at the feet of anthropogenic climate change, largely due to the infrequency and variety of events, and lack of historical records. Nevertheless, in recent years we have seen intense rainfall cause flooding, disrupt transport links, destroy property and infrastructure, and in some cases create public health risks and loss of life. It is important to understand that how urban areas cope with water is linked not only with the amount of precipitation but also the capacity of drains and rivers, the state of the tide, and atmospheric pressure. This was highlighted by the destruction of the village of Boscastle (Cornwall, UK) on 16th August 2004, which resulted from a series of factors which will be covered more generally later in this chapter. On that day, over 1000 trees were destroyed and 20 years' worth of river sediment was dumped in the village; 75 cars and 6 buildings were completely washed out to sea, and over 100 other buildings were severely damaged (figure 2.1). Luckily, no one died.

So, what caused this disaster? On 16th August, 185 mm (7″) of rain fell on the ground above Boscastle in the space of eight hours; the peak rainfall in Boscastle itself was recorded at 89 mm (3.5″) in an hour, more than a months' worth of rainfall! The rainfall was incredibly localised, and of the ten rainfall gauges in the area (all within a few miles), half showed less than 3 mm of rain that day. The local geography is significant: land rises steeply, and Boscastle lies in a valley with a natural harbour. There are over ten springs within a mile radius of the village, which all converge in the village, meaning there is always water flowing out to sea regardless of precipitation. To make matters worse, the flood coincided with high tide. Tides in August are relatively large, almost the size of spring tides. This coupled with the storm surge caused by the low-pressure system above the area, magnified the river flooding and prevented water flowing as quickly out to sea.

The torrential rain caused upstream rivers to bust their banks, causing 2 000 000 m^3 (2 billion litres) of water to flow down into the village; a full bath tub is around 1 m^3! The continuing rainfall caused the river estuary to rise by 2 m (6′6″) in one hour.

doi:10.1088/978-0-7503-1197-7ch2

Figure 2.1. A local shop after the flood. This was one of the few buildings left standing (This Boscastle flood damage photo has been obtained by the author from the Wikimedia website [https://commons.wikimedia.org/wiki/File:Boscastle_0011.jpg], where it is stated to have been released into the public domain. It is included within this article on that basis.).

A second flash flood was caused by cars, trees, and other debris becoming lodged under a bridge creating a dam, before breaking and releasing a second 3 m (10') wave. Unlike static flooding, this was dynamic, and water speed was over 4 m s^{-1} (10 mph, 14 km h^{-1}), more than enough to cause damage to buildings and wash cars out to sea (see figures 2.1 and 2.2). It has been estimated that more than 20 000 000 m^3 of water flowed through Boscastle that day. The cause was not entirely natural; while the rainfall amount and intensity was extreme, much of the damage was caused by anthropogenic activities. In addition to the cars and other debris becoming lodged under a bridge causing a second larger flash flood, changing in farming practices upland of the village with the removal of trees, hedges, and improved sewerage caused the water to flow much faster than it would have done in the past.

Such extreme events will be covered further in chapter 5. These disasters are linked to many contributing factors, many of which will be magnified as a result of climate change. The impacts for coastal or low-lying areas, or those adjacent to rivers, will be significant and are discussed in the rest of this chapter.

2.1 Sea level rise

Sea level rise is probably the most well-known and talked about aspect of climate change. While increasing temperatures at the poles promotes melting of the polar ice

Figure 2.2. View upstream after flood waters had subsided; note the damage to the building façade (This Boscastle flood damage photo has been obtained by the author from the Wikimedia website [https://commons.wikimedia.org/wiki/File:Boscastle_05.jpg], where it is stated to have been released into the public domain. It is included within this article on that basis.).

caps, this is not the main cause of sea level rise. In the same way an ice cube in a glass of water does not alter the liquid level in the glass as it melts, floating ice caps do not contribute to sea level rise. Land-locked ice masses and glaciers will, but this contribution is small compared to the primary cause, which is thermal expansion of the oceans as they warm. Due to the volume and dynamic nature of the oceans, they have a massive thermal inertia and will continue to warm and rise for several centuries after air temperatures have stabilised. This means that while we consider climate projections to the end of this century in this book, and some projections show stabilisation of air temperatures, the estimates of sea level rise are only the tip of the iceberg so to speak.

Sea level rise is a major issue for low-lying countries such as Bangladesh or Micronesia, where there is little infrastructure to hold back the sea. They could see large areas of their land mass become submerged. However, developed countries will also face significant challenges. Since much of The Netherlands is already below sea level (see figure 2.3), rivers have to be managed carefully so that water flows out to sea without extensive flooding. Therefore, extensive infrastructure is employed to hold back the sea and drain the land. Parts of The Netherlands are up to 7 m below sea level; as sea levels rise, dikes will have to be made taller, and pumps upgraded to be able to cope with greater amounts of water, to avoid extensive flooding. Due to the low-lying nature of the country, the Dutch have adopted the philosophy of living

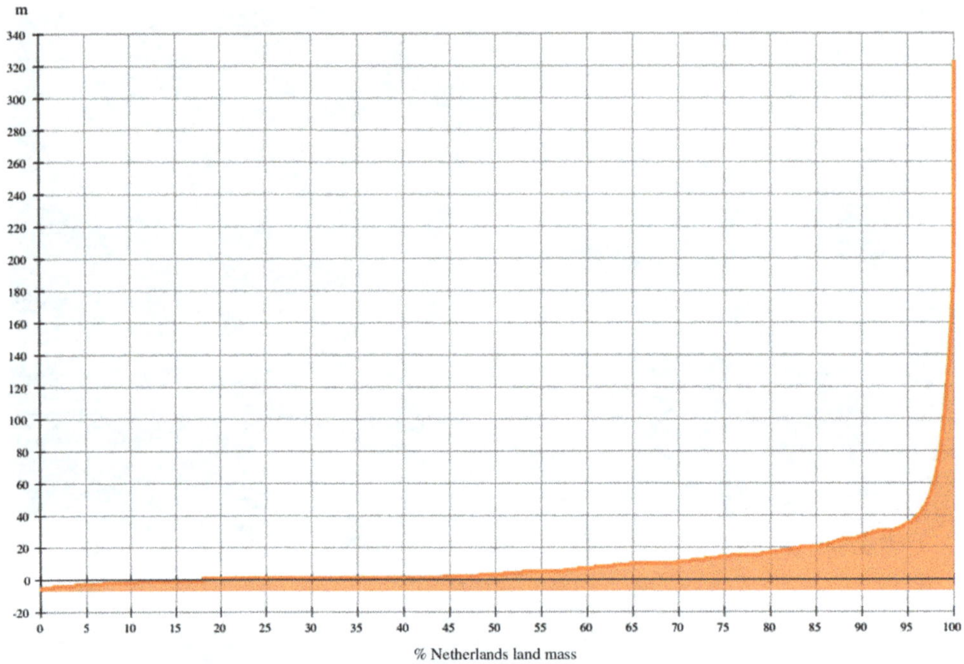

Figure 2.3. Cumulative distribution function of height above sea level for the Netherlands as percentage of land mass (This land height distribution image has been obtained by the author from the Wikimedia website where it was made available under a CC BY-SA 3.0 licence. It is included within this article on that basis. It is attributed to PawelS.).

with water rather than trying to control it as in other countries, which can be costly and very energy intensive. Hence, there is much other countries can learn about how to manage water in both urban and rural areas.

Other countries also employ measures to hold back the sea. The UK has amongst the largest tides in the world, and like many other European countries, its cities are built on, or are alongside, rivers and estuaries. The port city of London is no exception. Due to the topography along the Thames estuary, the Thames barrier, completed in 1982, was created to protect the main financial centre of the city and to protect over 200 000 properties in and around London. The barrier was originally designed to protect London from very high flood levels (up to 1 in 1000 year events) and to close if the tidal water level in central London were expected to exceed 4.87 m (16′) (a 1 in 100 year event). At the time of original conception (circa 1970) it was expected to be used 2–3 times per year with a useful lifetime of up to 2070, based upon the understanding of climate change and sea level rise at the time. Between completion in 1982 and March 2007 the barrier was raised 100 times to prevent flooding, greater than the initial estimates but still within expected limits. However, over the winter of 2012 the Thames barrier was raised 50 times, resulting in recommendations to the UK Government and Environment Agency that a new barrier be considered. The Thames Estuary 2100 plan considered options for the future protection of London and its environs from tidal flood risk for the rest of this century.

The Thames Estuary 2100 project found that by upgrading and modifying the existing Thames barrier, protection could be maintained until the 2070s. The strategy was designed to be able to cope if sea level rise was greater than the maximum stated in the IPCC 4th Assessment projections. A High-plus-plus (H++) scenario was devised to be used for such contingency planning. The H++ scenario is based upon extensive land-locked ice sheet and glacier melting, in addition to thermal expansion of the oceans, and hence represents a low probability, high impact scenario. The H++ scenario places an upper limit of a 2.5 m sea level rise during the 21st century, in addition to upper estimates for storm surges (covered later); this degree of sea level increase is feasible given current understandings, but is

Figure 2.4. Flooding of the Piazza San Marco Venice in 2012 (This Venice flooding photo has been obtained by the author from the Wikimedia website where it was made available under a CC BY-SA 3.0 licence. It is included within this article on that basis. It is attributed to Abxbay.).

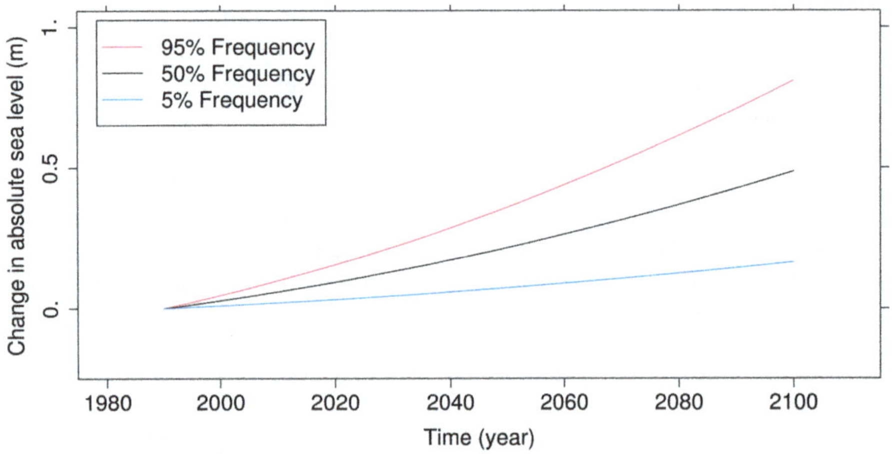

Figure 2.5. Estimates of average UK sea level rise over time for different (cumulative distribution function) probability levels (data: UKCP09).

considered very unlikely. Since the use of the H++ scenario in the Thames Estuary 2100 project, it has also been used for contingency planning for the provision of new sea defences at other locations around the UK.

A similar project is currently being built in Venice (Italy), to protect the historic city from rising sea levels and increasing storm surges. The MOSE project (*MOdulo Sperimentale Ellettomeccamico*), due to be completed in 2018, is a project intended to protect the city of Venice and the Venetian Lagoon from flooding. Similar to the Thames barrier, gates are raised to keep high tides and storm surges out. The low-lying nature of Venice means that large sections of the city are flooded if the defences are breached by even a small amount, as can be seen in figure 2.4. The MOSE system was deemed a better way to protect the heritage of the city than constructing sea walls, which would impact upon the cultural aspects of the historic city.

Since Venice experiences much smaller tides than the UK, the barrier has only been designed to protect against water heights of 3 m (10′), including 60 cm (2′) of sea level rise. To date, the highest tide (and storm surge) was 1.94 m (6′4″). Hence, the MOSE project might not be as resilient to future climate change as the Thames barrier; however, it covers a much larger area and spans some 18 km.

The current UK climate projections (UKCP09) provide estimates of possible sea level rise over this century. Figure 2.5 shows the range of likely sea level rise for the UK. As for other climatic variables, the projections of sea level rise are probabilistically represented by their relative cumulative distribution function frequency, i.e. the 5% level represents the value where 5% of projections are less than this value. While such information is not particularly easy to interpret at a glance, the probabilistic projections from UKCP09 allow a greater consideration of the risk of climate change.

2.2 Storm surges

The height of our seas and oceans is not static; it varies not only with the seasons and cycles of the moon affecting the heights of our tides, but also with the weather! A skew surge or storm surge is the height difference between a predicted astronomical high tide and the observed high tide. Storm surges occur as the name suggests, during storms or when a low atmospheric pressure weather system is above the ocean. High-pressure weather systems push down on the ocean's surface causing the water level to rise elsewhere, where there are low pressure systems. This can result in higher than expected high tides if the astronomical high tide corresponds with a low atmospheric pressure event. The lower the pressure, the higher the water rises; and as you might have guessed, the lowest pressure systems are often associated with storms, and often bringing intense rainfall and high winds. This can create an additional problem in that the higher than normal tide can prevent river water and surface runoff from intense rainfall from flowing out to sea, as was the case in Boscastle in 2004.

Storm surges can lead to extensive flooding, and are dangerous for people living in low-lying coastal areas. Sea levels rise roughly 1 cm for every millibar (mbar) decrease in pressure, and this can be further increased by the storm approaching the coast, the shape of the coastline and the seabed. For example, in 2005 when Hurricane Katrina approached the US coast, storm surges in excess of 8 m were generated in some areas, causing wide-spread damage to the city of New Orleans where the sea defences could not cope with the water level. Storm surges for the UK are typically more modest, ranging between 1 m to 3 m for different UK locations for a 1 in 50 year type event. Return periods such as this refer to the average amount of time between similar events. However, it is a mistake to think that if a 1 in 10 year event occurred this year we would not see a similar event for another ten years. A 1 in 10 year event could occur twice in the same year, or not at all for several decades. It is best to think of these return periods as the probability of an event occurring in any given year, regardless of when the last similar event was, i.e. 1 in 10 years would be 10%, 1 in 100 years would be 1%, and so on.

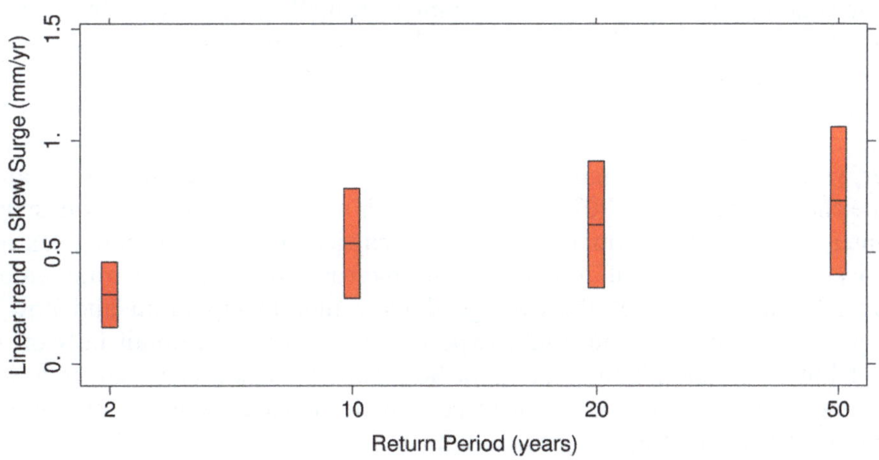

Figure 2.6. Expected change in storm surge height per year for different return periods. The boxes define the 5th, 50th, and 95th percentiles for the A1B emissions scenario; data shown for the Bristol Channel (data: UKCP09).

As the climate changes, storms are expected to become more intense, and the return periods of different events are expected to change. This will be covered more in chapter 5—Weather extremes. Additionally, reduction in the density of seawater due to thermal expansion (sea level rise) will lead to a relative increase in storm surge height over time. Figure 2.6 shows how the height of a storm surge associated with a given return period of storm may change over time in mm per year.

2.3 Flooding

Floods can have many causes. Non-coastal flooding can be split into three main types: fluvial (rivers), pluvial (surface flooding), and environmental (groundwater flooding). The most devastating floods are typically a result of rivers bursting their banks, particularly in urban areas where the impact on neighbouring property and infrastructure is immediate. However, periods of intense rainfall can cause surface water flooding and possible 'flash floods'. In this case, rainwater has not had time to be absorbed into the ground (percolation), either because of the surface type, as many urban surfaces are impermeable to water, or because the available drainage, whether ditches or sewers, are unable to cope with the volume of water. If the geology of the ground is permeable to water, then it is possible for flood water to come out of the ground if the water table raises high enough. Ground water flooding can be particularly damaging as it can be sustained for several months. This is because the water simply does not have anywhere to go; rivers and lakes will likely already be at capacity, and the ground is completely saturated.

The effects of climate change on rainfall are complex, and a warmer Earth will lead to greater evaporation of water from the oceans, which will lead to more rainfall. However, changes to global surface temperatures will affect global convection currents, and may well alter the path weather systems travel across the Earth (as discussed in chapter 1). For the UK, current estimates of climate change show very little deviation in the total amount of annual rainfall under any of the emissions scenarios (figure 2.7). Despite this, there will be a shift in when rainfall occurs.

Currently the UK climate is defined under the Köppen Geiger climate classification as CFB (temperate marine), implying we have a mild climate with no distinct dry or wet season. Climate change will see a shift in rainfall from summer to winter, with the amount ranging from effectively no shift at the lowest estimates of climate change under UKCP09 up to an ~70% shift in rainfall amount from summer to winter for the highest estimates. This poses particular risks for urban areas as the intensity of winter rainfall is expected to increase, not just the total amount. Figure 2.8 shows maps of the average daily winter precipitation and how this may vary over time. As you might expect, the majority of rainfall falls on high ground, but also predominantly along the South and West coasts. Urban areas in the South and West will have to adapt to cope with increased volumes of rainfall as current infrastructure may not be adequate.

Many historic cities are located near rivers or estuaries because the water way provided both food and transport allowing trade with other cities and countries. This required these water ways to be tamed; their routes were fixed, and are no long able to

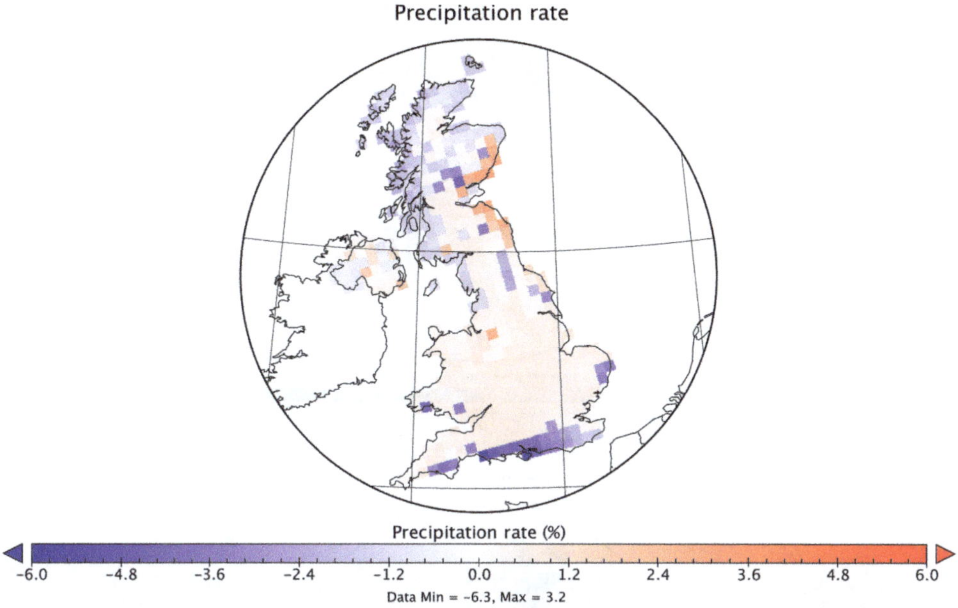

Figure 2.7. Relative change in annual precipitation by the end of the century (data: UKCP09 A1FI 50th percentile).

change according to geography and geology over time. While this allows a river to become a part of a city's infrastructure and a local amenity, there are draw backs. As the city grows, development encroaches upon land that would have previously been allowed to flood. Also, the river banks are sometimes reinforced with stone gabions, concrete, or sheet steel piling to prevent the river changing course. This has the desired effect of confining the river to a set course; however, it also confines the capacity of the river to a set volume of water, which if exceeded can cause considerable flooding. This has led to the construction of flood relief channels in some cities to provide additional capacity when the river exceeds a certain height, thus protecting property and infrastructure. Exeter is one such city, where a flood relief channel (figure 2.9) provides additional capacity, preventing the river Exe from flooding the nearby mainline train track and nearby roads in the St David's area of the city.

Like other rivers, the Exe has its source on high ground: in this case, Exmoor in the north of Devon. The photos shown in figure 2.9 show the river Exe's flood relief channel in the summer of 2016 (top) showing its typical water level, and again in late autumn (bottom), this time with an elevated water level covering the footpaths and part of the grass banks. You can see the spillway from the river at the end of the flood channel with the river beyond; normally, water is not flowing over this spillway, as shown in the top photo. The flood relief channel is approximately 3–4 m below the height of the river at this point (to the right of the bank in the photo), and re-joins the river approximately 1 mile downstream (1.6 km). You may notice from the photo that it is not raining; the only recent rainfall occurred on 19th November, and Exeter itself received minimal precipitation since it is located in the rain shadow of Dartmoor. Exmoor, however, received approximately 25 mm (1″) of rainfall on

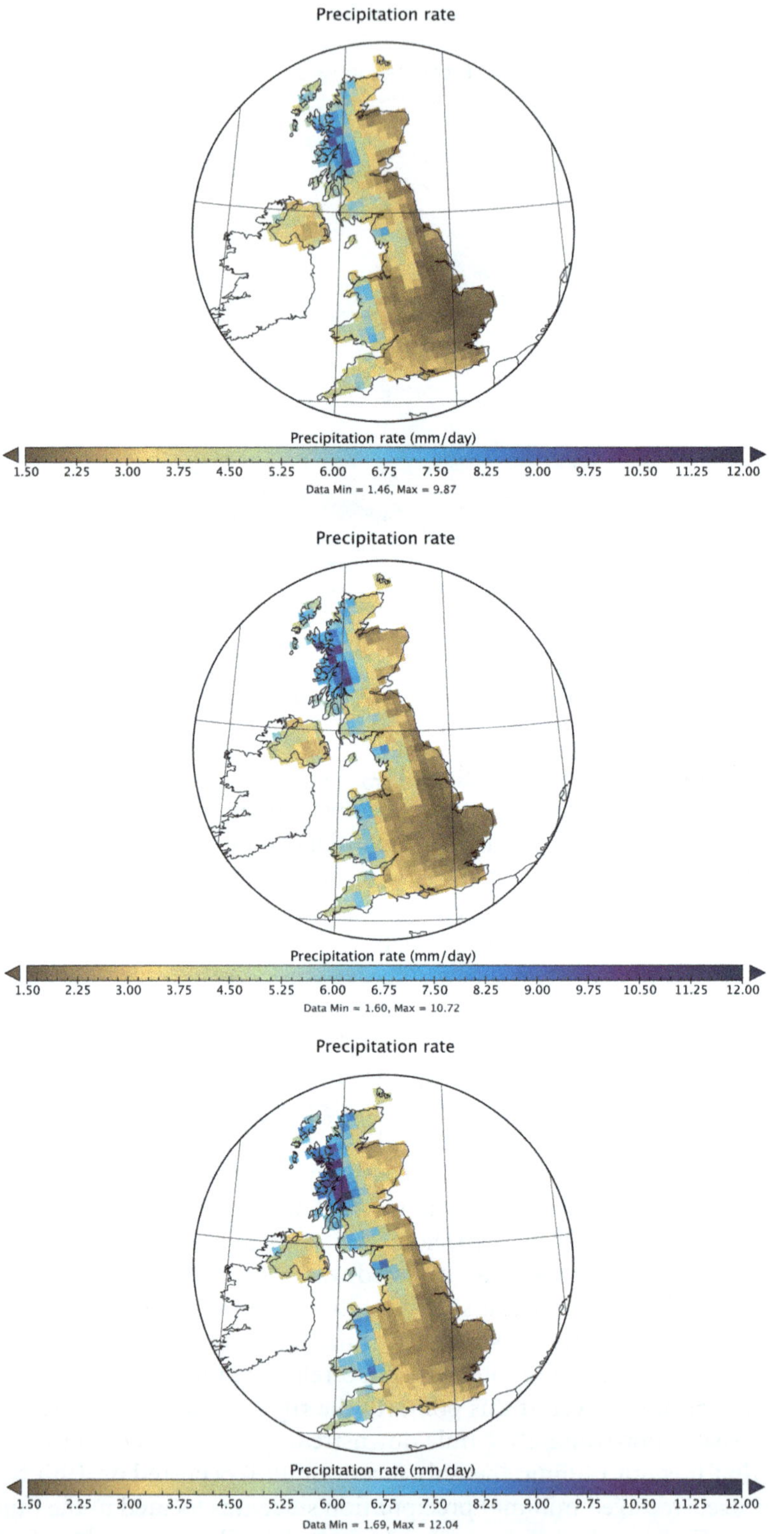

Figure 2.8. Average daily winter precipitation; central estimate for the 2020s (top), 2050s (middle), and 2080s (bottom) (data: UKCP09 A1FI 50th percentile).

Figure 2.9. River Exe flood relief channel 15th August 2016 (top), and 20th November 2016 following a day's heavy rainfall on Exmoor (bottom).

Figure 2.10. Road bridge over the river Exe on 20th November 2016.

that day. This highlights an important issue with urban rivers: it is not just the rainfall falling on the urban area and rapidly flowing into a river that cause flooding; there can also be delayed flooding from rainfall that occurred hours to days earlier upstream. Figure 2.10 is a photo of the river Exe adjacent to the flood relief channel, taken on 20th November 2016. There is approximately 30 cm (1′) clearance between the river and the road bridge at this point in time.

So long as the flood relief channel is able to take any excess water, this level will not rise further and the bridge is protected. However, this is dependent upon the flood relief channel having adequate capacity, not just in terms of total volume but also how much water can flow in and out per second. At the time of writing, the flood defences along this flood relief channel and further down the river are being raised to provide increased capacity and continued protection in the future.

The flood relief channel featured here serves its purpose but has to re-join the main watercourse at some point. Ideally, this would be outside of the urban area, but often this is not the case due to either geography or existing developments. In the case of the river Exe, the flood relief channel ends after approximately 1 mile (1.6 km) at the site of an old water mill, which is now a pub. On one side behind the pub, the bank rises steeply to the road and the city beyond; on the other side of the bank there is a grass bank followed by flat playing fields with houses beyond. The following two photos (figure 2.11) were taken from these playing fields showing how the river floods after a more prolonged period of heavy rainfall, in this case in the

Figure 2.11. The River Exe at the confluence of the river (right) and flood relief channel (left), as it looks normally (top) and during the floods of 2012 (bottom) (credit: top—author (12th October 2012), bottom—James Webb (25th November 2012)).

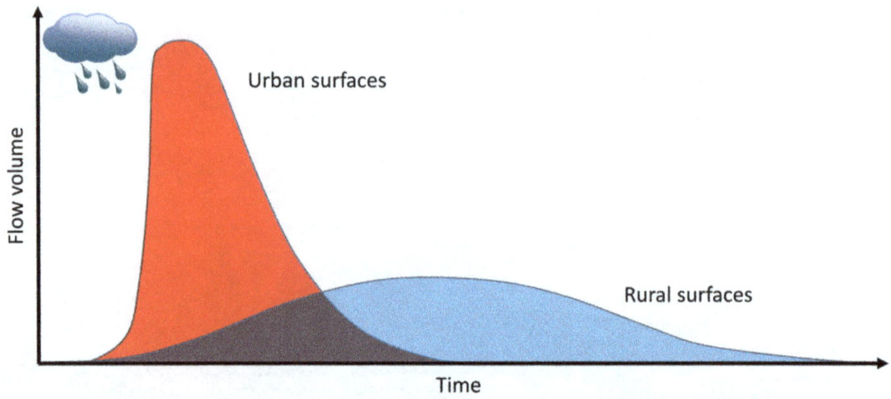

Figure 2.12. Illustrative hydrograph for typical urban and rural surfaces.

winter of 2012. As we can see from figure 2.11, the river at this point has risen significantly (>2 m), and the river basin (including footpaths and cycleways) is at capacity. This area had been designed to cope with such volumes of water and the level of protection is currently being increased; however, more historic areas such as Exeter Quay a little farther downstream, cannot cope as well. While recent apartment developments are raised above the river, the historic buildings (shops, restaurants, and tourist attractions like the old customs house) are low-lying, only ~1 m above the standard river height. The volume of flood water that passed through the Quay during the winter of 2012 caused significantly damage to both property and peoples' livelihoods. In this respect, for many cities like Exeter, it is the historic areas that are most vulnerable to flooding because they were not designed to cope with such large volumes of water, which result from upstream developments increasing peak water flow rates. The question then becomes how can we both protect and maintain our cities' heritage?

2.4 Flash flooding

Flash flooding can occur for several reasons both natural and anthropogenic. Flash flooding is different from normal flooding, which might be caused by prolonged rainfall saturating the ground and raising the water table to the extent that water does not soak in or drain away. Instead, flash flooding is most common after periods of intense rainfall, particularly when following periods of warm dry weather, where the ground becomes hard, compacted, and impermeable. The state of the ground is important as it is highly unlikely that enough rainfall will fall in a given location to cause flooding damage; instead, water is transported to, and converges in locations susceptible to flooding. This can be over the surface of the ground, either because of previous weather conditions or because of the presence of man-made impervious materials, such as roads.

Figure 2.12 shows an indicative hydrograph of the typical response for rain falling on an urban surface rather than a rural one. The area under each of the curves is the same, as the same volume of water is involved, but more water flows for a shorter amount of time over the hard, impermeable surfaces associated with an urban area.

Surprisingly, it is not just urban development that can increase peak flow rates and cause flash flooding. The events of 2004 in Boscastle discussed earlier in this chapter were attributed not only to intense rainfall but also clearing of scrub land uphill from the village. In this way, agricultural practices can also lead to increased flood risk: clearing of land, and removal of hedges, shrubs, trees, etc. can all increase peak water flow rates. All vegetation, whether trees, shrubs, of even grass slows water down through interception of falling rain and increased surface roughness. There is also the added benefit that rainfall intercepted by a plant canopy may never reach the ground; instead, it may be held long enough to evaporate directly from the plant canopy. Plants also use water from the ground, and this is passes as water vapour out through stomata in leaves in a process called transpiration. This helps prevent water saturation of the ground, and coupled with slower moving water, the likelihood of the water being absorbed into the ground through infiltration (also known as percolation) is increased. It is clear then that vegetation or land surface types that slow water down and give it time to be absorbed or evaporated are vital to help prevent flash flooding, and also help prevent pluvial and environmental flooding.

Climate change will not only shift year-round rainfall to predominantly winter rainfall, but also increase the intensity of rainfall events. Figure 2.13 shows how the amount of rainfall on the wettest day in winter will change by the end of the century

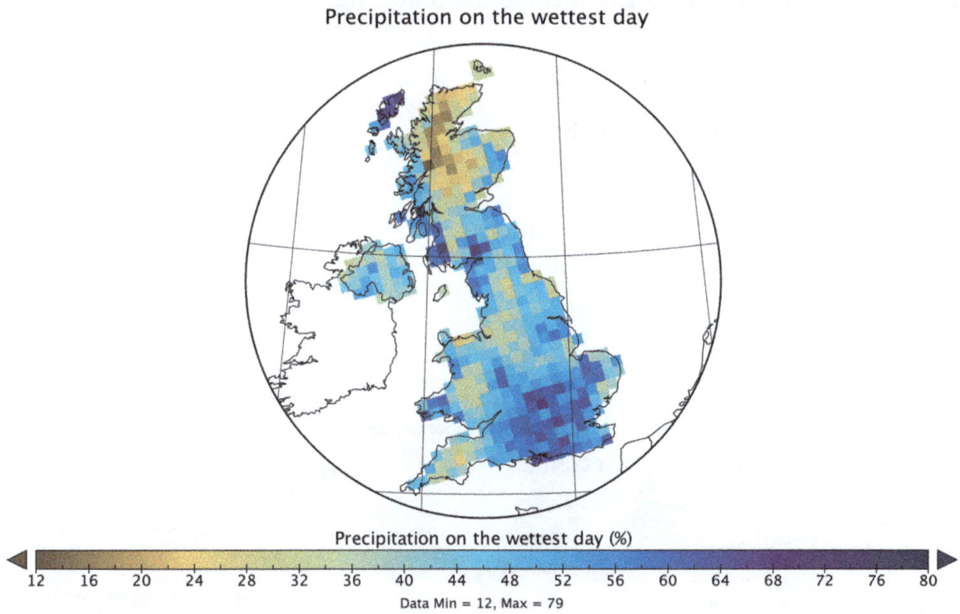

Figure 2.13. Change in the amount of precipitation on the wettest day in winter by the end of the century (data: UKCP09, A1FI 90th percentile).

under an upper estimate of climate change. High ground already receives a large amount of rainfall and this will not significantly increase; however, lower lying areas with large urban areas such as the South and East UK will see a large increase in intense rainfall. For example, the south coast around Portsmouth and Chichester which is already susceptible to flooding may well see a 79% increase in the amount of rainfall to fall on the wettest day in winter. Such intense rainfall will likely be accompanied by a large storm surge, which will amplify flooding risk for these coastal cities.

2.5 Potential solutions

Potential solutions to prevent or limit flooding in urban areas can be split into three different types: those that control water where it falls (source control), those that control water as it travels to collection points (site control), and those that involve wider regional control. The hierarchy of these different solutions is referred to as the urban water management train, as shown in figure 2.14.

Source control
Controlling water where it falls is the first step in limiting flood risk, and typically involves slowing water down so that that it has time to be absorbed into the ground. For most new developments, this takes the form of a Sustainable (Urban) Drainage System (SuDS) plan. A SuDS is a sequence of water management practices spanning both source and site control that aims to drain surface water in a more sustainable manner than simply routing water runoff to drains and through a pipe to a watercourse. An SuDS plan can include the following:

- Rainwater harvesting
- Green roofs
- Permeable surfaces
- Filter and infiltration trenches
- Swales
- Detention basins
- Bioretention systems or raingardens
- Retention basins, wetlands, and ponds.

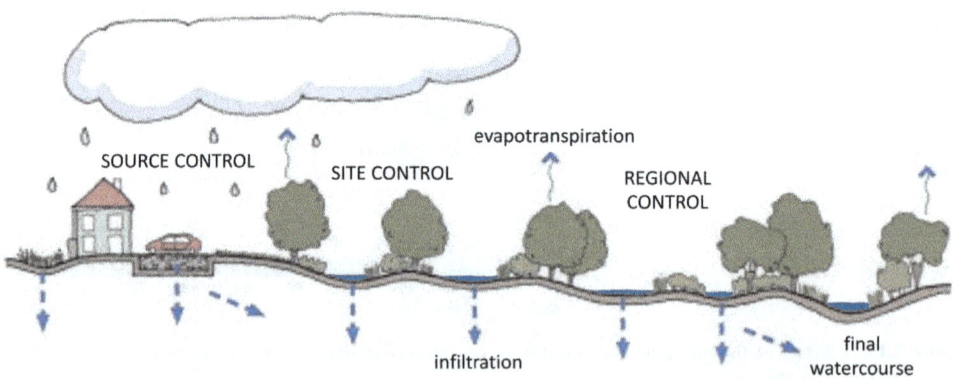

Figure 2.14. Urban water management train (adapted from SUSDRAIN).

Both collecting water from a roof (rainwater harvesting), and green roofs, such as the one shown in figure 2.15, slow rainwater and prevent it from flowing directly to a drain. In the case of rainwater harvesting, the water is collected in a tank and used for non-potable uses around the home, such as flushing toilets, washing clothes, or watering the garden. Water can be stored for days or weeks before it is used and enters drains or the ground. Green roofs only store water in the soil substrate, and this water is used by the plants and will have a net cooling effect on the roof and the surrounding air. These aspects will be covered later in chapter 4. In terms of controlling water, a green roof's primary purpose is to reduce water flow rates, both through collection and use by the plants. Furthermore, the rough surface presents a smaller steadier flow of water to a drain than say a tiled roof.

For other typically non-permeable areas such a car parks, pathways, and pavements, we can make the surface rougher to slow water down and use permeable materials and constructions to encourage percolation of water into the ground. Such permeable surfaces are now fairly typical for out-of-town developments, such as supermarket carparks built on an area that was historically flood plain. Figure 2.16 shows two examples of permeable paving suitable for carparks and pathways; however, other materials can be used, such as resin-bound aggregates, which are porous to water. The paving shown in figure 2.16 presents a rougher surface to the water than say tarmac, and hence water will flow slower over the surface. This relieves stress on drains, which have a finite capacity, and provides more time for water to pass through gaps between the pavers and to percolate into the soil below.

Figure 2.15. A sedum green roof atop the British Horse Society headquarters (This green roof photo has been obtained by the author from the Wikimedia website where it was made available under a CC BY-SA 4.0 licence. It is included within this article on that basis. It is attributed to Sky Garden.).

Figure 2.16. Examples of permeable paving (credit: SUSDRAIN).

Figure 2.17. Example of permeable paving coupled with water storage.

Such permeable surfaces can be coupled with a water storage tank, either as part of a rainwater harvesting system or just as an attenuation tank, providing storage before water flows to a watercourse or drain. Figure 2.17 shows an example of how this could be implemented: the paved road or carpark consists of pavers, a gravel substrate, and a water-permeable membrane resting upon plastic attenuation crates. The storage tank is made up of modular attenuation crates (figure 2.18) surrounded by a waterproof membrane. The crates are available with different loading capacities and can be used for purposes ranging from under a domestic patio or sports pitch, to a lorry loading bay. The modularity of the attenuation crates allows

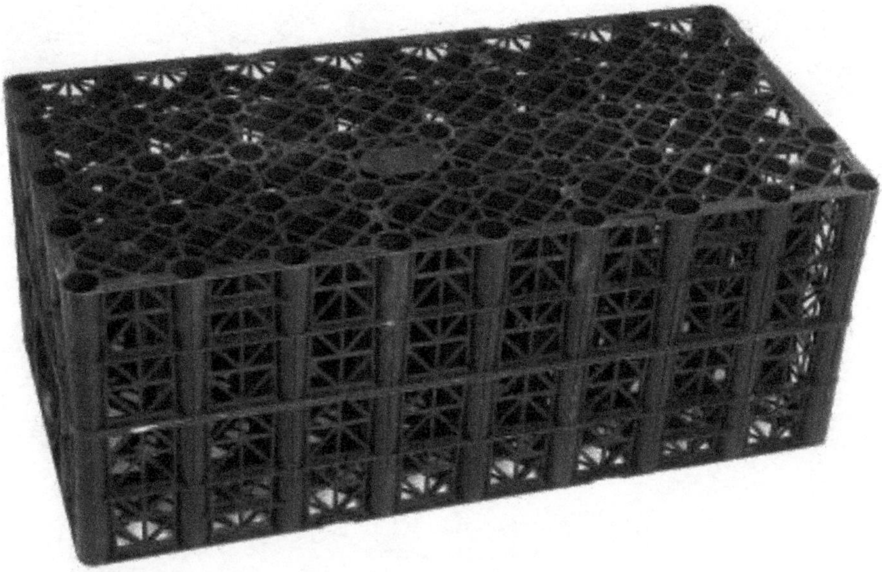

Figure 2.18. Image of a water attenuation crate for storm water runoff (credit: SUSDRAIN).

great flexibility in the shape and size of the water collection area and also volumetric capacity of such a system.

An alternative source/site control methodology to providing a permeable surface for water to pass through is to use conventional impermeable constructions and to guide water to an area where it can collect and drain away. A bioretention system or a raingarden such as the one shown in figure 2.19 is a shallow hole or depression containing vegetation that collects rainwater runoff from surrounding impermeable surfaces, such as roofs, driveways, roads, and pavement, providing the opportunity for the water to percolate into the ground.

Raingardens reduce surface runoff by allowing water to collect and be absorbed into the ground rather than flow into drains or across the surface. These can range from large areas such as the one shown in figure 2.19 to collect rainfall from a nearby road and pavement, to smaller areas collecting rainfall from a single roof or patio, replacing a soakaway. Raingardens also have the benefit of improving the water quality of nearby waterbodies by preventing pollution flowing into them and increasing groundwater for plants rather than sending stormwater straight out to sea.

Site control

Site control is an extension of the principles discussed above for source control, just on a slightly larger scale. This typically involves storing water runoff from a development or as an attenuation strategy at the culmination of a SuDS plan. Examples of site control include drainage ditches and swales (figure 2.20), which are simply depressions in the ground somewhere for water to collect and to drain away or be guided towards a suitable watercourse or drain. These have limited capacity and are typically used to collect water from a road or pathway.

Figure 2.19. A raingarden in Portland Oregon, designed to capture surface water from the road and pavement (courtesy of U.S. Environmental Protection Agency).

Figure 2.20. Swale drainage ditch.

On a larger scale, detention and retention basins provide somewhere for a larger volume of water to collect and to soak away, and can harbour many different species of flora and fauna, being a source of biodiversity for the site. The primary difference between a detention basin and a retention basin is the presence of water. A detention

Figure 2.21. A detention basin designed to capture rainfall from a nearby development.

basin (figure 2.21) simply holds water temporarily after a rainfall event and the water drains away over time and the basin is typically empty. A retention basin (figure 2.22) on the other hand, as the name suggests, retains water and could be considered a pond or reservoir with excess capacity.

Regional control

Regional control typically refers to the management of water courses and the management of flood plains to protect urban areas. One such example is the Somerset Levels in the southwest of the UK, which contains several watercourses, some natural and others manmade, with sluices to control water flow across the levels and out into the Bristol Channel. King's Sedgemoor Drain is one such manmade watercourse, which diverts the River Cary in Somerset to discharge into the River Parrett. As the name suggests, this watercourse was created between 1791 and 1795 to drain the local King's Sedgemoor peat moors, which were generally flooded, with the water only able to leave via evaporation. This was due to the existing drainage outfalls into the River Parrett being too high to be of much use. The 10.5 mile (17 km) channel of King's Sedgemoor Drain was completed in 1795, resulting in improved drainage to the surrounding area of 11 000 acres (4500 ha). The Drain has been upgraded over the last few centuries, being widened and deepened in places with the addition of tilting sluice gates to allow more flexible water level management.

Figure 2.22. A retention basin, capturing and storing water from a nearby development (This retention basin photo has been obtained by the author from the Wikimedia website where it was made available under a CC BY-SA 4.0 licence. It is included within this article on that basis. It is attributed to Drm310.).

Flood waters are removed from drainage ditches on the Somerset Levels by pumping stations; they were originally steam powered, but are now electric. The Drain itself operates entirely by gravity, this means that it is dependent upon the state of the tide in the Bristol Channel, which has one of the largest tidal ranges in the world. There were plans in 2002 to add a pumping station to the Drain, which would have allowed water to be pumped from the Drain into the estuary at all states of the tide, but these were not followed through. However, the presence of King's Sedgemoor Drain and its interface with other rivers such as the Sowy, Parrett, and Cary, along with numerous ditches, rhynes, and associated infrastructure, allows water to be managed on the Somerset Levels, including the selective flooding of certain areas, in order to protect the urban centres of Bridgewater and Taunton.

However, such extensive water infrastructure is not always enough, particularly in this case where a large collection area is dependent on the state of the tide to drain efficiently. The winter of 2013/14 is one such example, which saw the Somerset Levels flooded for several months. The combination of prolonged rainfall, large tides, and storm surge meant that the rivers and drains were at capacity, and water was only able to flow out to the estuary for a few hours each day before sluice gates had to be shut again to prevent tidal flooding. Thus, the prolonged rainfall saturated the ground leading to environmental flooding across the Levels, where water simply

Figure 2.23. Photo showing the confluence of the rivers Parrett and Tone during the flooding of February 2014 (This river Parrett flooding photo has been obtained by the author from the Wikimedia website where it was made available under a CC BY-SA 3.0 licence. It is included within this article on that basis. It is attributed to Rodw.).

rises out of the ground in low-lying areas rather than having to have flowed there, bypassing any flood defences. Figure 2.23 shows the result of the extensive rainfall over the winter and the wide spread flood across the Somerset Levels. In the photo, the two rivers, Tone and Parrett, are at capacity where they meet; however, the dykes are still visible above the rivers and flood waters. The flooding visible in the bottom left of the photo is a result of environmental flooding rather than of the rivers having burst their banks.

2.6 Conclusions

Given the likely changes to rainfall patterns across the UK as a result of climate change, flooding is going to become a more common occurrence. The integration of an SuDS plan is a valuable tool in combating flooding. Softening the landscape, adding permeable surfaces, green roofs, water butts, or rainwater harvesting, soakaways, and raingardens will all reduce the risk of flash flooding and by extension reduce the risk of pluvial and fluvial flooding.

However, such flood reducing strategies can be negated by both heritage and geography. If the urban landscape in question is historically significant, then altering it to incorporate an SuDS strategy, or even the widening and deepening of waterways or the creation of flood relief channels, is not possible. In the same way, if the urban area is surrounded by hills, like Boscastle, then the risk of flash flooding is greater because water will flow rapidly downhill and an SuDS plan may have little effect due to the volume and speed of the water. In this situation, maintaining hedgerows, copses of trees, and areas of scrub land on the hillside will slow water before it can reach the urban area. The issue is exacerbated further if the urban area in question is coastal and relatively low-lying, for example, the town of Fowey.

Fowey is located on the South Cornwall (UK) coast and has an undulating topography. This means that rain water will flow quickly into the valleys and streams (and the drains with Fowey itself) and other low-lying areas. The Environment Agency estimates that there are 290 properties, roads, a wastewater treatment plant, and two electricity substations at risk from a 1% probability (1 in 100 year) flood in Fowey. Over this century, it is estimated that the number of properties at risk will increase by around 14. The Environment Agency also estimates that there will be 20% more peak flow in watercourses based upon predictions of climate change, increased urbanisation, and changes in land use. This increases the likelihood of large-scale flooding in and around Fowey. Flooding of infrastructure, such as the water treatment plant and the two substations, can have disastrous effects and will likely be subject to increased flooding defences. This may not seem like a major issue, however, for a town like Fowey, much of the town's income is from tourism, based on the town's location and aesthetic (figure 2.24).

Fowey is a quaint historic Cornish seaside town, with narrow streets, which, due to the surrounding topography, have a steep incline (figure 2.25) down to the harbour and the main tourist centre. The streets are narrow in the historic part of the town, with no pavement, and are just wide enough for a car and a pedestrian. The streets are lined by stone walls, further confining where water can and cannot go.

The landscape levels off as you reach the town centre (Fore Street), which together with the promenade and marina constitute the main tourist centre of the town. Fore Street runs parallel to the estuary, with buildings backing onto the water, and several openings between buildings to allow access to jetties. There are currently minimal sea defences, and Fore Street is less than half a metre (20″) above the average high tide level (figure 2.26).

The low-lying nature of Fore Street and the main tourist centre of Fowey means flooding is inevitable each time there is a particularly high tide, whether astronomical or due to a storm surge. This also means that sea level rise is a particularly

Figure 2.24. Panoramic view of the Fowey Estuary, which is home to many yachts and pleasure craft during the summer months.

Figure 2.25. Three streets in Fowey. Note the steepness and width of the streets; water has nowhere to go but downhill.

Figure 2.26. View of Fowey sea front and harbour wall from the water; you can see the mean high tide level on the walls.

Figure 2.27. Images of Fore Street when flooded by spring tides (March 2008) and as it appears normally (July 2016). Flooding images courtesy of Fowey Renewable Energy Enterprise (FREE).

pertinent issue for the residents of Fowey. The median estimate of sea level rise by the end of the century is approximately 0.5 m (figure 2.5), which is sufficient for water to overcome the current sea wall with almost daily regularity. Figure 2.27 shows Fore street full of tourists on a sunny summer's day and also during a spring high tide, with sand bags outside shop doors to help limit flooding.

As the events at Boscastle in August 2004 showed us, storm surges and heavy rainfall can coincide with disastrous consequences. While an entire month's rain falling in such a short period of time is certainly extreme, such events may become more common in the future. Furthermore, the coincidence of these two events is quite likely since a low-pressure system capable of creating a large storm surge typically produces a lot of rain. This can exacerbate these individual climate impacts since the rainwater within rivers and streams will not be able to flow out to sea as readily due to the storm surge. Within Fowey there are implications for storm drains, particularly for the low-lying area around Fore Street, which already suffers during storm surge events. Water can enter drains uphill, but exit again at the bottom of the hill due to increased resistance from the storm surge preventing this water exiting into the estuary.

For towns such as Fowey, adaptation to cope with the combination of sea level rise, greater storm surges, and more intense rainfall is unappealing. The H++ scenario used by the Environment Agency for a sea defences such as the Thames Barrier sets an upper limit on sea level rise of 2.5 m by 2100. This conservative

Figure 2.28. Photo showing an opening from Fore street to moorings and jetties on the estuary. Currently there are no sea defences and water is free to flow through to the town during particularly high tides.

Figure 2.29. Photo showing the same area as figure 2.28 during a high spring tide (March 2008). Photo courtesy of Fowey Renewable Energy Enterprise (FREE).

scenario is likely to be used for sea defences around critical infrastructure such as electricity sub stations and water treatment plants, to protect them from sea level rise and increased storm surges. Such visually intrusive flood defences will impair the local aesthetic and have an impact on tourism. While increasing temperatures may well increase tourism numbers in the shorter term (sea levels rise slower than temperatures), a long term strategy to protect or even move the town centre will need to be employed. At the very least, openings from Fore Street to the estuary, such as the one shown in figure 2.28, will need to be protected to prevent occurrences such as the one shown in figure 2.29.

Fowey is not alone in this respect, and this issue is not limited to small coastal villages and towns; many of our historic cities are built on rivers and tidal estuaries. Such flooding events are the entire reason for flood defence infrastructure such as the MOSE project and the Thames Barrier, and we are likely to see other similar defences deployed to protect our cities and cultural heritage sites in the near future.

IOP Publishing

Climate Change Resilience in the Urban Environment

Tristan Kershaw

Chapter 3

Temperatures

People often talk about global warming when referring to climate change. However, the term global warming is avoided by climate scientists for the reason that not all places on Earth will get warmer; some may get cooler as a result of changing global circulation patterns. For instance, if the Gulf stream were to move, the UK would experience a very different climate, being on a similar latitude to Moscow (Russia) and Alberta (Canada). That said, for the vast majority of urban areas it is highly likely that temperatures will increase as a result of climate change.

This poses a problem because the way we design our buildings and urban areas, and even our way of life, is determined by the background climatic conditions, particularly air temperatures. Buildings have two main purposes: to provide shelter from the elements, and to provide thermal comfort. The latter is determined by the human metabolism and the local environmental conditions. Our bodies produce heat as a by-product of the biological processes going on in our bodies. As warm-blooded creatures, we maintain a core body temperature, meaning that sometimes we have to lose heat or keep heat in depending on the local environment; generally, the easier it is for our bodies to maintain this core temperature, the more comfortable we feel. This has fed into how we design our buildings. In the UK and other Northern countries, buildings are predominantly designed to keep heat in, while in the tropics, buildings are designed to keep heat out. As the climate changes, our buildings and the way we use them will have to adapt to continue to provide both shelter from the elements and thermal comfort.

The press is full of estimates of changing temperatures due to climate change, such as a 2 °C increase, or a +4 °C world. However, these figures are a global average. The oceans' ability to store large amounts of heat means that they warm slower than land masses; thus, a global 2 °C increase is closer to 4 °C–5 °C in London (UK). Also, some areas will warm faster or slower than others due to latitude, altitude, or proximity to the sea. Figure 3.1 shows the output of a global climate model (GCM) used for the IPCC 5th Assessment Report. The figure shows

doi:10.1088/978-0-7503-1197-7ch3

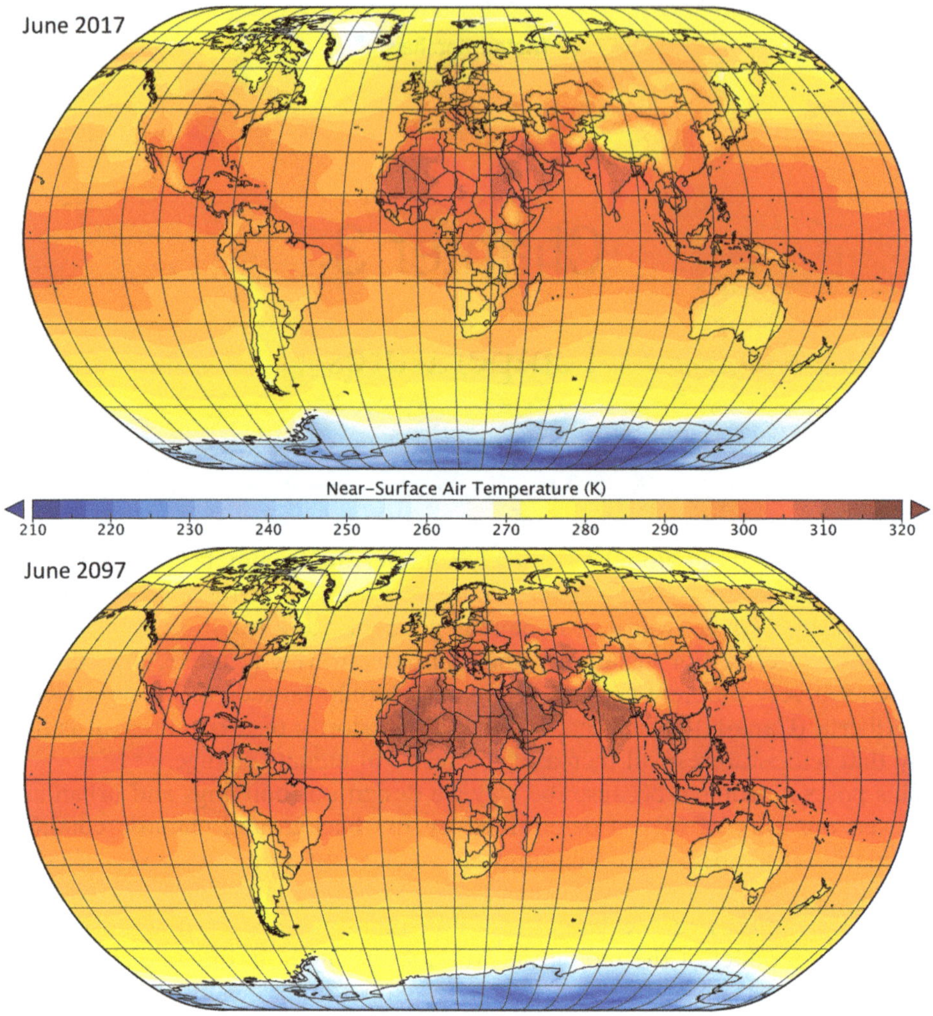

Figure 3.1. Images of near surface mean air temperature for the current climate and the end of the century under a high emissions pathway (data: HadGem2 RCP8.5).

average near surface air temperatures for the month of June, for 2017 and 2097. Since the scale is in Kelvin, subtract 272 to get degrees Celsius.

We can see from figure 3.1 that over the 80 year period spanning the two images that the Earth has gotten warmer; however, the warming is not uniform, either due to topography (e.g. Himalayas) or albedo (Antarctica). While RCP8.5 is a higher estimate of climate change (see chapter 1), the consequences of this level of climate change are severe. The temperatures reached in areas of Africa and Asia for example are high enough to cause serious heat-related illness to human beings. If we look in more detail at the UK using regional climate model (RCM) data (figure 3.2), we can see how temperatures are projected to change over the course of this century. We can

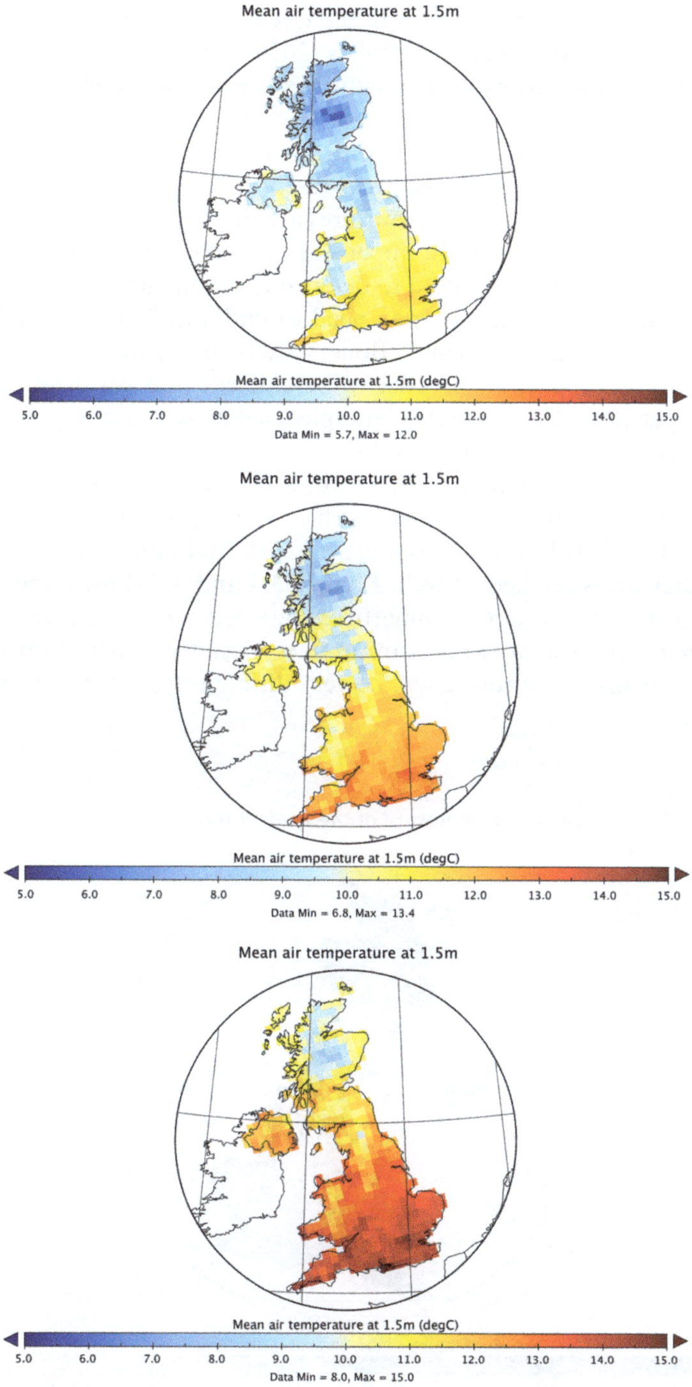

Figure 3.2. Mean annual air temperature for the UK for the 2020s (top), 2050s (middle), and 2080s (bottom) (data: UKCP09 A1FI 50th percentile).

see that mean annual air temperatures in the UK can expect roughly a 4 °C increase between the 2020s (2010–39) and the 2080s (2070–99).

Despite these changes to mean temperatures across the globe and in the UK, at the lower end minimum temperatures are not expected to change dramatically. Figure 3.3 shows how the temperature of the coldest night in winter may change by the end of the century (2080s). Most areas see little rise in minimum temperatures, particularly coastal areas where temperatures are moderated by the sea. This means that climate change will have little impact on the way we specify heating systems since the peak load temperature used to determine the capacity of, say, a boiler, will not change much. This implies we may need to find a different way of heating our homes in order to maintain efficiency. Boilers are typically inefficient if only used at part of their overall capacity. In order to be efficient and yet meet the peak load demands of a particularly cold night, multiple smaller systems may need to become the norm rather than a single large boiler.

The small change in minimum temperatures implies that the maximum temperatures must increase considerably to achieve the predicted changes in mean temperatures. The distributions of daily, seasonal, and annual temperatures is not just shifted; they are stretched as well. Figures 3.4 and 3.5 show upper estimates of the change to wintertime and summertime maximum temperatures, respectively. With changes in winter temperatures of up to 6 °C and summer temperatures of up to 12 °C, the seasons of the future will be very different from those we remember or experience currently.

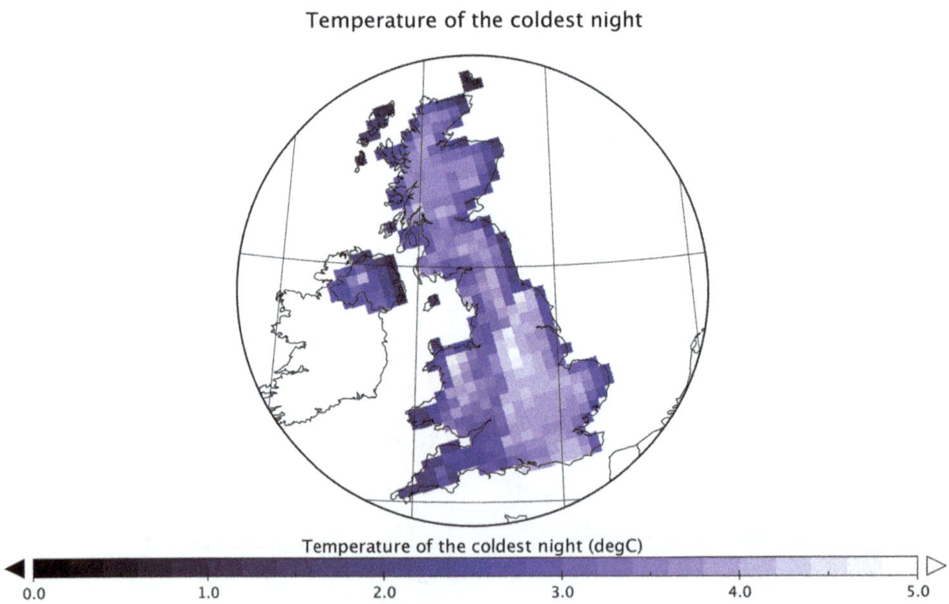

Figure 3.3. Change in temperature of the coldest night by the end of the century (data: UKCP09 A1FI 50th percentile).

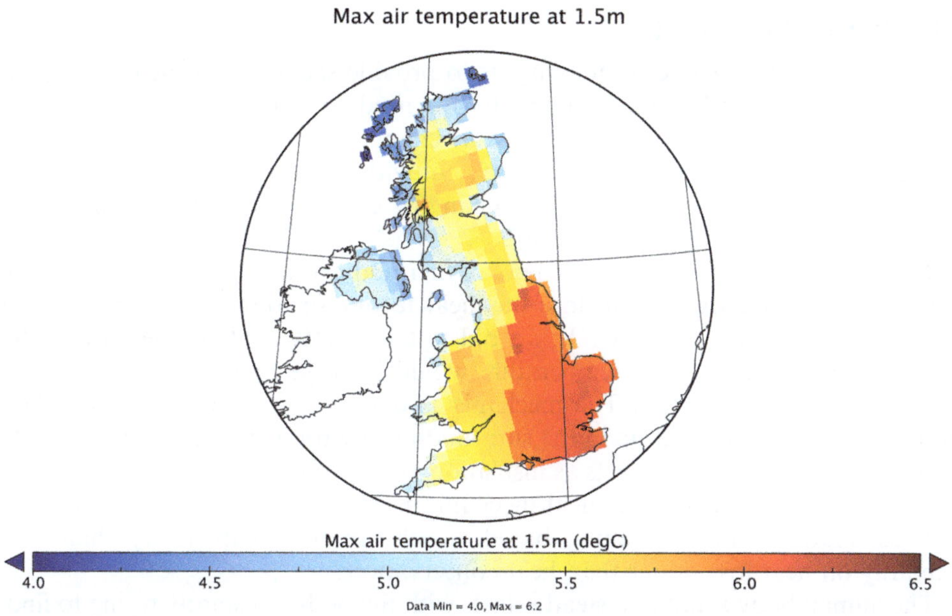

Figure 3.4. Change in average winter maximum temperatures by the end of the century (data: UKCP09 A1FI 90th percentile).

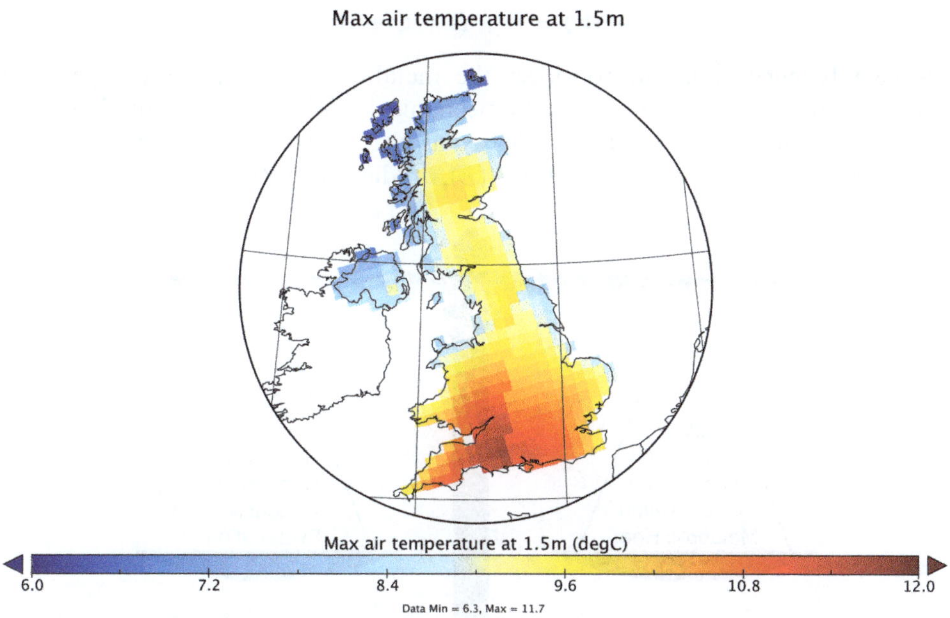

Figure 3.5. Change in average summer maximum temperatures by the end of the century (data: UKCP09 A1FI 90th percentile).

3.1 Human physiology

Since the primary purpose of buildings is to provide shelter from the elements and thermal comfort for building occupants, we need to consider how this may be impacted by climate change. We will discuss changes to the design and construction of buildings to resist weather in chapter 5, but with regards to how to maintain the thermal comfort of the occupants in a changing climate, we first need to consider what drives our perception of comfort.

The human body is subject to the laws of thermodynamics; we generate heat from energy in food and we need to lose this heat to our surroundings to maintain an internal temperature of 37 °C. If we did not lose this heat to the surrounding environment, our core body temperature would increase by about 1 °C per hour, leading to death soon after. The mechanisms our body uses to control our body temperature are dependent upon the surrounding environmental conditions and the clothes being worn; clothing acts as insulation around our bodies, limiting heat loss. Our body has several physiological mechanisms to control our body temperature, such as dilating and contracting blood vessels. In this chapter, we shall focus primarily on heat loss rather than heat conservation.

The human body is not in a steady state, with the body constantly trying to find a balance between heat gain and heat loss depending upon the level of activity and the environmental conditions. We can introduce a conceptual heat balance equation for this dynamic system:

$$(M - W) = E + R + C + K + S.$$

This heat balance equation considers the metabolic rate (M) of the person in question, which produces energy to do mechanical work (W) (e.g. climbing stairs), the remainder of which ($M - W$) is released as heat. This is balanced by heat lost or gained via evaporation/condensation (E), radiation (R), convection (C), or

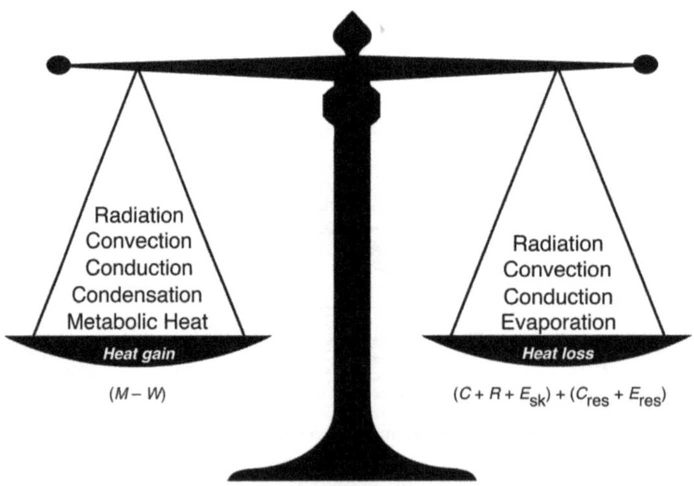

Figure 3.6. The conceptual heat balance. If both sides are equal, then a core body temperature can be maintained.

conduction (K). If the heat gains and losses are in equilibrium (figure 3.6), then a core body temperature can be maintained; however, if gains are greater than losses, theat is stored in the body (S) and the core body temperature increases (or falls if losses > gains).

The conceptual heat balance described above is a useful starting point, but is not much use if we actually want to consider the environmental impact on human physiology. In order to make this useful, we need to consider the specific values for heat production and heat exchange for a human body. Estimates of metabolic heat production are typically provided in Met for different activities. 1 Met = 50 kcal m^{-2} h^{-1} = 58.15 W m^{-2} (surface area of the body) and is representative of the heat output of a seated person. Further estimates of the metabolic rate for different activities are presented in the table below:

Activity	Estimate of metabolic rate (W m^{-2})
Lying down/sleeping	45
Sitting	58
Standing	65
Writing/typing	85
Walking on a flat surface at 2 km h^{-1}	110
Painting	130
Laying bricks	150
Carrying a 10 kg load while walking on a flat surface at 4 km h^{-1}	185
Walking on a flat surface at 5 km h^{-1}	200
Loading a wheelbarrow with rocks	275
Cycling on a flat surface at 20 km h^{-1}	360
Going up stairs	440
Running on a flat surface at 15 km h^{-1}	550

In order to relate this to a specific individual, we can calculate the surface area of a person from their height and weight using the Du Bois surface area (A_D):

$$A_D = 0.202W^{0.425}H^{0.725}$$

where W is the weight of the person (kg), and H is their height (m). Typically, values of 1.8 m^2 and 70 kg are used for men, and 1.6 m^2 and 55 kg for women. The surface area of the person is necessary in order to calculate the available area for heat exchange with the surroundings and in turn how comfortable or uncomfortable we are doing certain activities in different conditions.

The Danish scientist Povl Fanger carried out extensive work on the human perception of thermal comfort and how this is related to metabolism and environmental conditions. Detailed in his 1970 book '*Thermal Comfort: Analysis and Applications in Environmental Engineering*', his work is still the primary metric for assessing comfort nearly 50 years later. He used a modified version of the heat balance equation:

$$M - W - E_{dif} - E_{sw} - E_{res} - C_{res} = R + C$$

where

E_{dif} = heat loss by water vapour diffusion through the skin;
E_{sw} = sweat rate;
E_{res} = latent respiration heat loss;
C_{res} = sensible (dry) respiration heat loss.

This can further be rearranged into the form shown in figure 3.6:

$$M - W = (C + R + E_{sk}) + (C_{res} + E_{res})$$

where E_{sk} is the total evaporative heat loss through the skin (both sweating and vapour diffusion). In this equation we have two distinct heat loss paths: via the skin and via our breath. This is a useful way to view this heat balance as it makes it clear how we can impact different heat loss paths. Wearing more clothes for example insulates our body. It reduces our outer surface temperature (the temperature of the clothing in this case, not the skin), thus reducing heat loss by radiation and convection; however, it also reduces the ability for our skin to lose heat by diffusion or sweating. Increased clothing however, will have no effect on heat loss via our breath. Fanger went on to quantify how each of the variables in our heat balance equation is affected by different environmental conditions, such as air temperature, the temperature of surrounding surfaces (radiant temperature), humidity, air velocity, amount of clothing, posture, and level of activity.

Thermal comfort is often defined as 'that condition of mind which expresses satisfaction with the thermal environment'. Since no-one has ever gone into a room and thought this room is (thermally) comfortable, we should consider thermal comfort as a lack of noticeable discomfort. A person's perception of what makes them uncomfortable is highly complex, depending on individual metabolism, clothing, age, weight, and ethic origins. It is also highly subjective. The reasons why a person reports thermal comfort (or discomfort), or related reports of warmth, freshness, stimulation, or pleasure, are not well understood. There is no single solution to determine if someone will be comfortable or not, and it is widely accepted that you cannot please everyone all the time. Instead, the emphasis is on what environmental conditions will produce thermal comfort for the majority, based upon six basic parameters: air temperature (t_a), radiant temperature (t_r), air velocity (v), humidity (rh), clothing level (clo), and metabolic heat (M–W).

Fanger developed a system for quantifying how satisfied or dissatisfied a person is with their local environment, based upon the rated thermal sensations of a large group of test subjects. The subsequent metric predicted mean vote (PMV) is still widely used, and is featured in many building codes worldwide to identify suitable internal environmental conditions in buildings. PMV is ranked on a 7-point scale, from −3 to +3, as shown below.

Hot	+3
Warm	+2
Slight warm	+1
Neutral	0
Slightly cool	−1
Cool	−2
Cold	−3

The calculation of PMV considers all six basic parameters mentioned above, along with air pressure, in order to calculate sweat evaporation from the skin, and convective and radiative heat loss. The methodology is complex, and requires not only the insulative properties of the clothing worn but also the percentage coverage of the skin. This allows the area for sweat evaporation to be calculated, along with the temperatures of the skin and the clothing respectively. Calculation of PMV is now programmed into many building energy/thermal analysis software packages as a metric to assess building performance. Where this calculation fails however, is that it can not estimate discomfort from temperature gradients, e.g. a hot head and cold feet, which may arise from temperature stratification in a room. Despite this, PMV forms the basis for many other calculations (e.g. productivity covered later) and other metrics, such as the predicted percentage dissatisfied (PPD).

$$\mathrm{PPD} = 100 - 95\exp(-0.03353\mathrm{PMV}^4 - 0.2179\mathrm{PMV}^2).$$

This related estimate of thermal discomfort is interesting in that it ranges from 5%–100%, indicating that 1 person in 20 will always be dissatisfied with the local environment.

As you might expect, the most powerful mechanisms we have to control our body temperature are behavioural. We can add or remove clothing, change our posture to increase or decrease the effective surface area for heat exchange, open windows or fan ourselves, move to shade, and so on. Our bodies also have a number of physiological mechanisms to help maintain our core temperature. Both our behavioural and physiological responses continually interact at a subconscious level to respond to changing environmental conditions. These physiological mechanisms include, amongst others, vasodilation and vasoconstriction of skin arterioles, i.e. to increase/decrease blood flow close to the skin to increase/decrease radiative and convective heat loss. Sweating increases evaporative heat loss at the skin's surface; this coupled with vasodilation can assist in cooling the blood and hence the body core. In cold conditions, we shiver to heat ourselves. Shivering effectively increases our metabolic rate from a rest rate of 65 W m^{-2} to around 200 W m^{-2} or more. All these different responses work to thermally regulate our bodies, primarily to safe guard our health, but also to provide comfort. This thermoregulation however, has an impact on our mental focus and levels of productivity, due to fidgeting, distraction, complaining, time off work, etc.

Currently, there are no comprehensive policies for how to adapt existing homes, offices, schools, hospitals, or care homes to higher temperatures. Both average and maximum temperatures are expected to increase across the UK over the next few decades. Previous extreme events such as the European heat wave of August 2003 are expected to be representative of a typical summer by the 2040s. Such events have been shown to have a significant impact on the productivity of workers. The heatwave of 2003 is estimated to have cost the UK between £400–500 million in reduced manufacturing output over the two week period.

There have been several studies of how environmental variables such as temperature affect the productivity of workers. It is generally accepted that there is economic benefit for work places to maximise worker satisfaction. The question is what factors increase worker satisfaction in the work place? There is evidence that giving people personal control over temperature and ventilation improves perception of comfort and increases satisfaction. In hot environments, blood vessels in the skin will dilate to increase radiative and convective heat loss, then increase sweat rate to increase latent heat loss; if these are insufficient the core body temperature will rise. Increasing core body temperature raises heart rate and causes fatigue, since the body must work harder to try and maintain its internal temperature. The resulting heat stress and discomfort will lead to behavioural changes and effects on cognitive performance, such as mental performance, information processing, memory, and so on.

Many worker tasks require both physical and cognitive functions. For example, typing requires speed and accuracy while processing information. Being in thermal discomfort affects different aspects of tasks in different ways, and thus it is necessary to consider a metric of thermal comfort such as PMV that accounts for all these factors and the level of activity being undertaken. Many office workers use computers for a significant part of the day. Computing is a special task requiring close attention to details and visual cues. The requirements of computers for the user to follow specific logic-based procedures results in a high optimal level of arousal.

Moderate heat stress of a few degrees Celsius above optimum can have an adverse effect on the performance of tasks due to relaxation and a reduction in arousal. This is considered to be an autonomic response to regulate body temperature at the limit of vasodilatory control, prior to the onset of sweating. It has been noted that a conscious effort can override this effect and maintain arousal levels. Entering a hot room and being suddenly exposed to high temperatures can act as the que to exert this conscious effort; however, in most cases temperatures will rise slowly over the course of the day, circumventing the opportunity to exert conscious effort, and resulting in lower arousal and concentration. Worker performance depends upon the level of arousal compared to that required for optimum performance. A task that is considered boring will be de-arousing, and a subject will perform better if their arousal level can be raised by some means. Stimulation caused by heat, cold, or draughts may increase arousal, but can also lead to distraction and time off task if arousal is increased too much.

If we are to consider how adverse thermal environments will affect worker performance, we need a relationship between a comfort metric, such as PMV, and worker performance. Roelofsen [1] compiled a direct relation between loss of performance and thermal comfort based upon several studies of worker performance and models of how humans respond to thermal load:

$$P = b_0 + b_1\mathrm{PMV} + b_2\mathrm{PMV}^2 + b_3\mathrm{PMV}^3 + b_4\mathrm{PMV}^4 + b_5\mathrm{PMV}^5 + b_6\mathrm{PMV}^6$$

where P is the loss of performance as a percentage, and the values $b_{0\text{–}6}$ are regression coefficients, whose value depends on whether you are on the hot or cold side of neutral. Roelofsen's relationship between productivity and PMV can be seen in figure 3.7. The red data point indicates thermal neutrality (PMV = 0), we can see that there are maxima in relative productivity on either side of neutrality, with the greatest being on the cool side. This is to be expected since, as previously discussed, an increase in arousal due to slight discomfort will improve productivity.

In addition to reduced productivity, increased temperatures and the resultant heat stress from trying to maintain a core body temperature can have health implications. If heat needs to be dissipated, then the body's central blood volume decreases so that blood vessels near the skin can dilate (increasing volume). This reduces the stroke volume of the heart, which must then beat faster in order to maintain blood volume flow rate to the bodies organs. The effective circulatory blood volume will further decrease after the onset of sweating as water is lost from the body. If this water is not replaced, then the heart must work harder (as well as faster) to pump the more viscous blood around the body. If sweating persists for a prolonged period, the body can become dehydrated, which can lead to further physiological stress and illness.

The impacts of hot weather events, such as heat waves and warming from climate change resulting in overheating in public buildings such as hospitals, care homes,

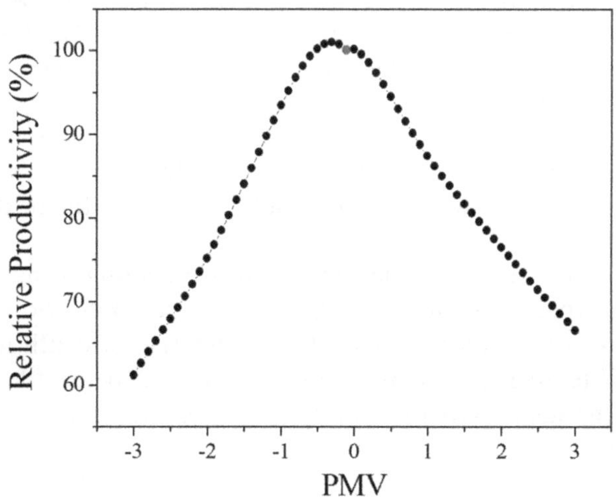

Figure 3.7. Plot of the relationship between productivity and thermal comfort.

schools, and prisons across the UK, are not currently known. This is because high temperatures can have greater impact on the very young and elderly whose bodies are unable to regulate temperatures as effectively. Moreover, infirm and vulnerable groups are particularly susceptible, and it is currently not well understood how different conditions, illnesses, or even medications affect our body's ability to control temperatures.

Even for healthy, able people who can take behavioural steps to cool themselves and whose bodies are able to deploy the full range of physiological responses to high temperatures, there is still danger from prolonged exposure to high temperatures. It is currently unclear how quickly humans will adapt to a changing climate. If we are unable to adapt quickly enough, it is estimated that the annual UK heat-related mortality (currently around 2000 deaths per year) will increase by two-thirds by the 2020s, increasing to ~250% by the 2050s, and by more than 500% by the end of the century. These estimates are based not only on a warming climate (a medium emissions scenario (A1B) was used) but also on a growing, ageing population. The percentage of people in the UK aged over 75 is expected to increase from 8% in 2015 to 18% by 2085. More elderly people are more likely to be adversely affected by high temperatures, due to factors such as reduced sweating efficiency; however, at temperatures above 35 °C, all age groups are at risk. It is possible that humans will undergo some form of autonomous physiological acclimatisation to gradual increases in mean temperature. For instance, it is possible for someone to emigrate from one climate to another and over time become acclimatised to that climate. However, evidence suggests that people are unable to adapt to sudden changes in temperature over a short period of time, particularly if overall temperature variability also increases, as is expected as a result of climate change.

Worryingly, there is evidence that people lack a basic understanding of the risks to health from high indoor temperatures, and as a result are less likely to take measures to safeguard their and any dependents' wellbeing. The UK Committee on Climate Change warns that insulating homes to improve winter thermal efficiency needs to be undertaken carefully to avoid increasing the risk of overheating. This is largely due to the way that we currently design our homes, and is an indication that the climate change mitigation and adaptation agendas are not always complimentary. This is something that will need to be addressed over the next few decades to safeguard the population.

We can estimate the risk to human life from overheating using the summertime two day mean external temperature ($T_{\text{2day}}^{\text{mean}}$), which has been shown to be strongly associated with excess summer deaths. The relative risk of death increases linearly above a threshold temperature, coincident with the 93rd percentile of $T_{\text{2day}}^{\text{mean}}$. Thus, we can estimate the heat-related mortality (M) over the summer from the relative risk (RR) given by

$$M = D_{\text{summer}}^{\text{all-cause}} \times (RR - 1)$$

with

$$D_{\text{summer}}^{\text{all-cause}} = \frac{d}{365} \times D_{\text{year}}^{\text{all-cause}}$$

and

$$RR = \left(\alpha \times T_{\text{2day}}^{\text{mean}}\right) + \beta$$

where $D_{\text{summer}}^{\text{all-cause}}$ is the deaths over the summer from all causes, d is the number of days when the $T_{\text{2day}}^{\text{mean}}$ is above the threshold temperature, and α is the heat-mortality gradient in % per degree above the mortality threshold temperature (a measure of how adapted people are to higher temperatures). External air temperatures are not always the best indicator for assessing risk to human health; instead, temperatures inside buildings are more appropriate. It follows then, that the internal mortality threshold temperature would be the 93rd percentile of internal temperatures. This assessment allows a comparison between different types and constructions of buildings and their resilience to higher temperatures. Liu *et al* [2] mapped the overheating risk and related summertime mortality for different building types and constructions across a medium-large city in the UK (Sheffield, population ~553 000), for the 2020s (2010–39) and the 2050s (2040–69). The external mortality threshold temperature for Sheffield has been shown to be 22.2 °C with α equal to 1.7%. β can be calculated when RR equals 1 and $T_{\text{2day}}^{\text{mean}}$ is equal to the mortality threshold temperature.

It is currently unclear at what rate people will adapt to the higher temperatures expected in the future. If people do not adapt quickly, the mortality threshold temperature for the 2050s will be similar to the 2020s; however, if people do adapt then it follows that the mortality threshold for the 2050s will be the 93rd percentile of 2050s temperatures, and any increase in heat-related deaths will be due to changes in the shape of the temperature probability distribution function. Maximum temperatures are expected to increase more than minimum and mean temperatures, implying that the distribution will become wider and more skewed.

As stated above, the heat-related mortality M is determined from the linear relationship between relative risk (RR) and the internal two-day mean temperature ($T_{\text{2day}}^{\text{mean}}$) above a citywide mortality threshold temperature. The citywide mortality thresholds for Sheffield were calculated as 24.4 °C and 25.9 °C for the 2020s and 25.4 °C and 27.2 °C for the 2050s for the 50th and 90th percentiles, respectively. Figure 3.8 shows the results of Liu *et al*'s [2] simulations showing mortality maps of Sheffield at a ward level. Variation in mortality rate between wards is due to differing altitudes and different building constructions.

Average projections of citywide M for the 2020s, at the 50th and 90th percentiles are 7 and 12 per million per year respectively, increasing to 21 and 39 per million per

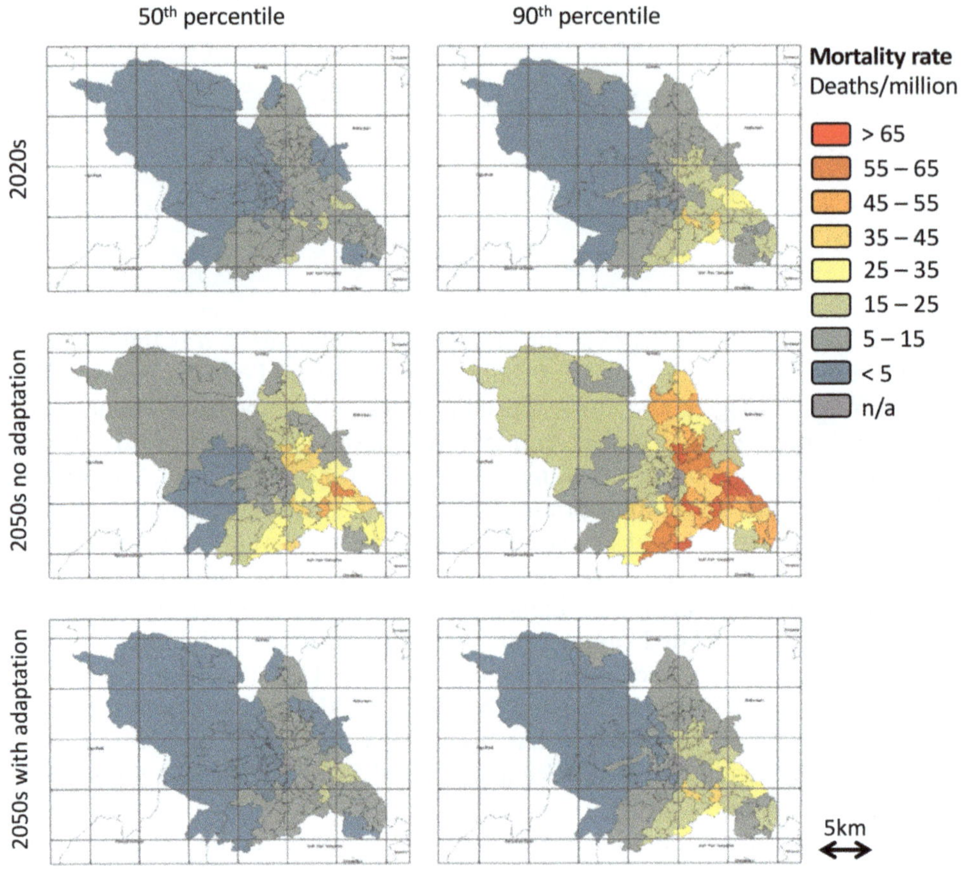

Figure 3.8. Map of heat-related mortality at a ward level for Sheffield for the 2020s and 2050s for different levels of climate change (credit: Liu *et al* [2]).

year in the 2050s in the absence of any adaptation (using the 2020s mortality threshold temperature). This indicates that the heat-related mortality rate would roughly triple in a 30-year period if people were unable to adapt to a warming climate. This is in line with estimates from the committee on climate change. If people do adapt quickly enough to increasing temperatures (using a 93rd percentile threshold temperature for the 2050s), the adverse impact of the warming climate on M is largely offset by human physiological adaptation. The mortality rates shown in figure 3.8 are comparable with those recorded in other UK cities during hot periods of weather. What is clear from the work of Lui *et al* is that the type and construction of the dwelling is important. It was found that terrace and semi-detached dwellings showed the greatest amount of overheating, particularly at night. This was due to the combination of mainly solid brick walls with poor insulation (which can keep heat out as well as in), with the high thermal capacity material (bricks) exposed to direct solar radiation. This allows the bricks to reach a high temperature and the heat to enter the building relatively easily. Finally, both terrace and semi-detached homes have a reduced surface area over which to lose heat by radiation and convection due to the party walls between dwellings, and detached houses showed comparatively lower overheating risk despite similar building constructions. Interestingly, bungalows showed the highest daytime temperatures but the lowest level of nocturnal overheating in the study. This was attributed to the large roof area compared to the volume of the building. Since all rooms are on a single floor in a bungalow, the area of roof exposed to solar radiation is much larger than for a two storey (or greater) dwelling. Therefore, there is a large area to absorb heat from the Sun, which is several hundred W m^{-2} depending on time of day, azimuth, and inclination of the roof. However, this also means that there is a large surface area over which to lose heat at night, meaning that the bungalows cooled faster and showed less nocturnal overheating. It seems clear then that there is the potential to adapt the way we design our buildings and urban areas to reduce heat storage and overheating, thereby reducing the risks to human health and increasing task productivity.

3.2 Building physics and possible adaptations

Overheating in buildings is clearly a pressing issue. However, there is evidence that as we attempt to make our buildings more energy efficient by increasing levels of insulation and standards of air tightness, we are also increasing the risk of overheating. The National House Building Council (NHBC) has collated several sources of evidence regarding the construction of new homes, the refurbishment of old ones, and the risk of overheating. Historically, our homes have been designed to keep heat in to provide thermal comfort in winter, with only small openings for ventilation and no solar shading. Although we have made our buildings better at keeping heat in and simultaneously increased the amount of heat generated in our homes, but we have not fundamentally changed the way we design or use our buildings, and thus overheating seems inevitable. Our homes are smaller than ever before and we have more heat-generating equipment within them, such as more TVs, computers, tablets, phones, set top boxes, home entertainment systems, etc. All of these generate heat, and yet we have not increased provisions to remove this heat from our buildings in the summer months. We also have not added solar shading to reduce solar radiation entering through windows in the summer months. Double glazed windows are now the norm in the UK; each pane of glass only reduces the amount of solar radiation (visible and near-infrared) entering the building by a few percent, but reduces the heat (thermal infrared) that can pass back out again by about half (glass is opaque in the thermal infrared). Without shading, windows placed on façades to provide the light airy rooms that we favour in the UK can easily lead to high temperatures and heat that cannot escape from our buildings due to high levels of insulation and poor ventilation.

The materials used to construct our buildings and urban areas are extremely important because they are what control the exchange of heat between the internal environment and the outside world, via radiation, convection, and conduction. The emissivity ε (a measure of how absorbing or reflective an object is) of the materials used is especially important in sunny locations. Radiation striking an opaque surface can be reflected or absorbed (usually a mixture of the two) depending upon its ε. The colour of a surface gives a good indication of the ε in the visible spectrum. The lighter the colour, the more reflective with a lower ε, while darker surfaces are more absorbing with an ε closer to 1. Hence, darker surfaces absorb more solar radiation and will attain higher temperatures than lower ε surfaces. The ε of a surface is also a measure of how radiative a surface is at different wavelengths, with the most absorbing surfaces ($\varepsilon \approx 1$) also being the most radiative. However, the colour of a material tells us nothing about the ε of the material in non-visible wavelengths of light, such as the near-infrared or thermal infrared. Paint, for example, can be tailored in the visible spectrum to different colours, and hence different emissivities; however, due to the carbon in the paint, all colours are highly absorbing/radiative in the thermal infrared. Hence, both white and black painted surfaces emit heat equally at night even if their absorption of solar radiation differs during the day. Thus, light coloured paints have been used for centuries in warmer locations to help keep buildings cool. The Italian island of Elba (figure 3.9) is a prime example of this phenomenon; the buildings covering the island come in a wide variety of colours, but all are light in hue with a low ε in the solar spectrum.

The buildings of Elba and of many other Mediterranean countries employ light coloured painted façades to reflect the solar radiation away from the building. This however, is not enough to maintain thermal comfort. Buildings in warmer climates are typically made of heavyweight materials, which can absorb heat from the air during the day to be lost at night when air temperatures are lower.

Heat storage within a material depends upon its mass m (kg) and its specific heat capacity c_p (kJ kg^{-1} K^{-1}). The ability of a material to store heat within a building is

Figure 3.9. The many coloured buildings of Elba (Italy).

often termed thermal mass, and the greater the specific heat capacity or the greater the mass, the more heat Q can be stored:

$$Q = c_p m \Delta T$$

where ΔT in this case is the difference in temperature between the material and the surrounding air. The specific heat capacity c_p for many typical building materials are relatively similar, even between those used as insulation and those considered lossy. For a material to have good thermal mass it needs to be relatively conductive to heat to allow it to maintain a uniform temperature; otherwise, the surface would simply warm up to meet the air temperature and heat transfer would cease. This also imposes the limitation that the mass should be relatively thin, else heat would not have time to travel through the material to access all available mass. This implies that the best thermal masses are dense materials so that there is substantial mass that can be accessed by heat travelling by conduction from the surface. Thermal mass can be quantified by the volumetric heat capacity of a material (kJ m^{-3} K^{-1}), a measure of how much heat can be stored per unit volume. Materials such as stone and concrete are good examples of thermal mass.

The greater the thermal mass of a building, the greater the effect it has on moderating the external air temperatures. The high specific heat capacity and density of thermal mass means a large amount of heat can be stored in the material, which implies that the surface temperatures will be lower than a low thermal mass material. A lower surface temperature reduces heat lost by the material through convection and radiation. Reducing the rate of heat loss means that the heat stored within the thermal mass takes longer to be released, shifting the diurnal temperature cycle and offsetting peak temperatures. Figure 3.10 shows an example of internal air temperatures for both a thermally lightweight building (e.g. timber frame or aerated concrete blockwork structure) and a thermally heavyweight structure (e.g. cast concrete, brick or blockwork) compared to the external air temperature. There is still a lag in time and a reduction in the peak temperature for the lightweight building due to the insulative properties of the structure, but these are minimal compared to the effects of the heavyweight structure. With a thermally heavyweight building, internal temperatures can be significantly reduced by several °C and the extremal temperatures offset by several hours. This can be particularly useful, say, for an office building where significant heat is generated during the day from people and equipment as well as any solar gains. A heavyweight structure will allow this heat to be stored in the structure to be released after the working day has ended, keeping the internal environment cool during working hours, reducing or even avoiding the need for air conditioning.

Part of the reason why newly built homes are at risk of overheating, in addition to improved insulation and air tightness, is the choice of materials and the construction methods now used. There is now a prevalence in the UK to construct new buildings out of materials that are quick and easy to use, thereby saving time and money. Where once internal walls would have been brick, now plasterboard and timber stud work is common. Meanwhile, brick and block outer walls have been replaced with lightweight timber frame or aerated concrete block walls. Perhaps the biggest change

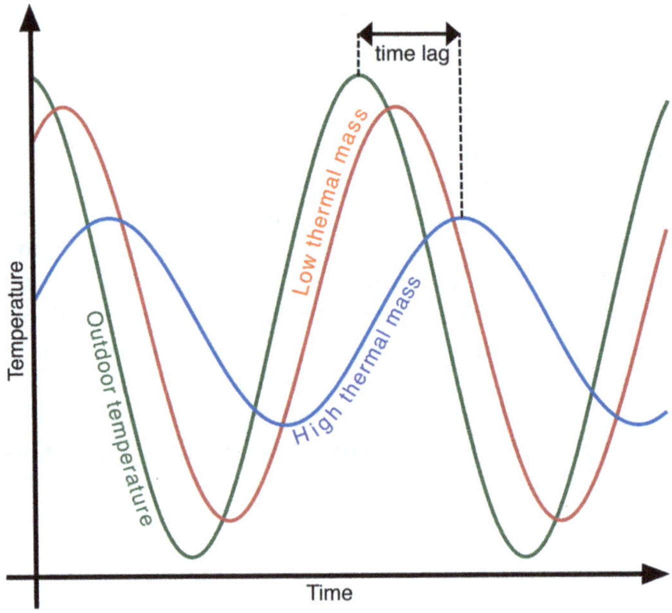

Figure 3.10. The relationship between time and temperature for different building constructions.

is the move away from wet plaster finishes to plasterboard being used instead. Unlike solid plaster, plasterboard is more insulative and is not in direct thermal contact with the wall construction behind it due to a small air gap. This has the effect of masking any thermal mass that may be in the wall, making the building act as a thermally lightweight structure even if heavyweight materials have predominantly been used. A similar effect can be seen in modern office buildings, which are often highly glazed, meaning that the majority of the thermal mass is in the concrete floor slabs between storeys. However, on one side this is covered with carpet or another insulative floor finish, while the underside of the concrete floor slab is masked by a suspended ceiling to hide services and improve room acoustics. However, limiting thermal mass or masking it behind an insulative material can be beneficial in some cases. The presence of a significant amount of thermal mass makes the building very slow to respond to changes in environmental conditions. This also means that thermally massive buildings are slow to warm up when the heating comes on or to cool down when a window is opened or air conditioning is turned on. Hence, there is a trade-off between peak temperatures and responsiveness to building controls, which needs to be considered when designing a building.

Buildings are not just stand-alone entities; they exist in thermal balance with their surroundings. Hence, when considering how to improve the thermal performance of a building we also need to consider the surrounding context. As we discussed in chapter 1, solar radiation heats the Earth's surface. The resultant temperature depends upon the latitude, season, time of day, the inclination of the ground, and the type of terrain (figure 3.11). Natural vegetative cover of terrain moderates extreme temperatures and stabilises conditions through evapotranspiration, increased albedo

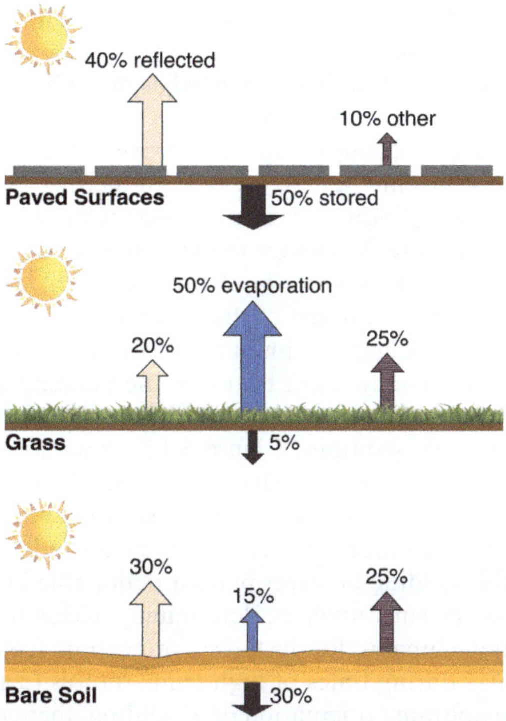

Figure 3.11. Heat balance for different surface types, showing proportion of reflected solar radiation, heat lost by evaporation or radiation/convection/conduction (shown as other) and heat stored in the ground.

(reflectivity) and shading of the ground below. Plants and grassy terrain cover tend to reduce surface temperatures, but cities and man-made surfaces tend to increase surface temperatures. This will be covered further in chapter 4. An indicative heat balance for different surface types is shown in figure 3.11. A common mistake that can lead to unpleasant environmental conditions is to place paved surfaces, which store a large amount of heat due to their density and specific heat capacity, close to the windows of buildings. These paved areas not only have the drawbacks discussed earlier in chapter 2, but also reflect a significant proportion of solar radiation, which can then enter buildings via windows. Roughly half the solar radiative power is converted into heat, warming the paved surface. Due to the thermal mass of the paving, a large amount of heat can be stored, meaning that the paving will still be hot and radiating a large amount of heat after air temperatures have fallen. This can have an adverse effect in summer evenings when windows are open and we are trying to purge our dwellings of heat, since the paved surface will not only heat the local air but also radiate heat directly into our building, thus preventing it from cooling effectively. By comparison, grass is a much cooler surface due to a large proportion of the solar power going into the evapotranspiration of water from the soil surface and through the stomata of the blades of grass. Even bare soil, due to its permeable nature, allows for solar power to be used for the evaporation of moisture from the soil's surface. This is true for other permeable surfaces as well, manifesting as a reduced surface temperature (figure 3.12), providing an additional benefit for the SuDs systems discussed in chapter 2.

Our experience of heat in a building is affected as much by the heat radiated from the surfaces around us (the radiant temperature) as it is by the air temperature. Therefore, our perception of comfort can be improved in hot conditions by having cool surfaces around us; this implies the use of high thermal mass materials to keep surface temperatures down. If you walk into a stone building, such as a church, on a hot day you immediately sense the change in the radiative environment. The human body is almost a perfect black body in the thermal infrared spectrum, and hence is very sensitive to changes in the amount of thermal radiation received. The colour of the stone is also important. Lighter colours reflect more of the solar spectrum than dark materials, reducing the amount of heat that the building absorbs. A dark wall exposed to the Sun can have a surface temperature 10 °C greater than a light coloured wall in the same conditions. Figure 3.12 shows some indicative temperatures for different surfaces around an urban area exposed to sunlight. The choice of materials deployed in our buildings and urban areas can have a profound impact on our perception of thermal comfort by altering the mean radiant temperature.

The orientation of a building or street plays a major role in the thermal balance between a building and its surroundings, determining access to wind and receipt of solar radiation. In hot climates for instance, protection from solar radiation is paramount, particularly during times of high temperatures (e.g. just after midday). When considering the optimum orientation of a building, there needs to be a balance between minimising solar receipt during the hottest months and solar receipt during the coolest months. Generally speaking, east- and west-facing walls receive the most radiation on the vertical surface due to the low solar angles at the beginning and end of the day. Keeping east/west-facing walls shorter with few openings is beneficial in preventing heat from entering the building. Openings on west-facing façades can be particularly problematic as they receive solar radiation during the warmest part of the day (the land surface has had time to warm), and the low solar angles can penetrate into a building, causing substantial overheating. As climate changes, there is a need to consider how best to alter the design and construction of our buildings to

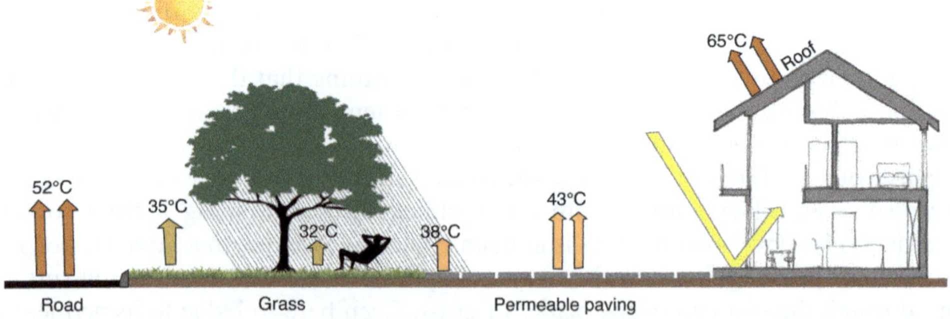

Figure 3.12. Indicative temperatures of different surfaces around buildings. Temperatures are indicative of a sunny day in a warm dry climate.

ensure that they continue to provide shelter and thermal comfort to their occupants. Luckily, by comparing projections of future climates with those found already elsewhere around the globe, this does not need to be a process of trial and error.

3.3 Learning from other architectures

Civilisation has evolved over the last few millennia during a period of relatively constant climate, allowing primitive architecture, defined as buildings of pre-literate societies whose knowledge of architecture and construction came from word-of-mouth, to attain high levels of performance against environmental stimuli. This implies knowledge of the performance of building materials and the effects of climate. Vernacular architecture, referring to architecture native to a particular context or location, which has been informed by factors such as tradition, environment, and social norms, can be considered as the evolved form of primitive architecture. Hence, vernacular architecture, urban layout, and culture (the way we use buildings) have evolved to be resilient to local climate conditions, and to provide both shelter and thermal comfort. For example, architecture at lower latitudes have evolved to minimise the urban heat island (discussed further in chapter 4) and to prevent buildings overheating under intense sunlight. Likewise, buildings in the tropics have to be resistant to high winds and driving monsoon rains. Climate change is set to alter the status quo, changing the way our buildings will need to perform. By examining what architectural responses have evolved elsewhere we can draw inspiration of how to adapt our building designs to the impacts of climate change.

The shape of a building has historically been informed by the local climate of the region in question, either to keep heat in or to keep heat out. Victor Olgyay in his book '*Design with Climate*' expresses the opinion that the optimum shape for a building is one elongated in an east–west direction. This allows minimum solar receipt in summer on the east and west façades but greater solar access in winter on the south façade (in the Northern Hemisphere). In hot, arid regions however, there is still a great risk of overheating in summer, and thus the building becomes more inward looking, with interior courtyards. Courtyards have been a dominant feature in Middle Eastern architecture for centuries. In the town of Ur (Iraq), excavated foundations have shown that courtyards were incorporated into building design as far back as 2000BC. It seems then that humans have been using architectural responses to high temperatures for millennia. Over time these have migrated along with people to other locations where they have been adapted to different climates and cultures.

Southern Spain, is often thought of as a haven for European tourists along the Costa del Sol; however, north of the coast are mountain ranges and plains that receive the most intense sunlight of anywhere in Europe. For the majority of the year, these plains enjoy a relatively mild climate, but during the summer temperatures can regularly exceed 40 °C. Thus, the local architecture over the millennia has evolved to not only keep this intense heat out and safeguard the building occupants, but also to provide thermal comfort during the hottest periods of the year. Lessons in the application of heat-sensitive design can be found in the palaces of the region. The Alhambra located in Granada is a complex of both palaces and fortresses dating originally from 889 as a small fortress built upon Roman fortifications. It was converted to a royal palace in 1333 for Yusuf I, Sultan of Granada. However, after the Christian Reconquista in 1492, ending ~770 years of Islamic rule of the Iberian Peninsula, the Alhambra became the site of the Royal Court of Ferdinand and Isabella, and was updated to better reflect the current Renaissance styles. It was further updated and added to by Charles V, Holy Roman Emperor, who added a further Renaissance palace. The Alhambra and other regional examples, such as the Royal Palace of Seville, are a fusion of Romanesque and Islamic architectural styles, termed Mudéjar, which flourished in the period following the defeat of the Moors and the return of Christian rule to this region of Southern Spain. The Mudéjar style did not involve the creation of new shapes of building like the Gothic movement, but rather an incorporation of Islamic art and architecture into the building forms of existing medieval Christian architecture.

Central to the palaces that make up the Alhambra are decorative courtyards such as the Court of the Lions (figure 3.13). Courtyards are an extension of interior rooms, accessed by covered walkways, defined by the differentiation of three different spaces: the exposed centre, the transitional edge, and the protected building interior. Of these spaces the transitional space offers an attractive living space during the heat of the day, well-lit by diffuse and reflected sunlight but shaded from the direct heat of the Sun. Unlike the building exterior, which contains few openings, the walls bordering the covered walkways and galleries contain many openings onto the courtyard beyond. The centre of the courtyard is not intended as an occupied space as might be expected in courtyard gardens in the UK; instead, this is home only to plants and fountains. The interior of the building beyond the covered

Figure 3.13. The Court of the Lions within the Alhambra (Granada), note the fountain as the central feature of the courtyard and the extended eaves and covered walkways surrounding the courtyard (This Court of the Lions image has been obtained by the author from the Wikimedia website where it was made available under a CC BY-SA 2.0 licence. It is included within this article on that basis. It is attributed to comakut.).

walkways and galleries are cool and dark. This is a significant departure from the way that we currently use buildings in the UK, where we expect our building interiors to be well lit.

Shading is the most effective defence against the heat of the Sun and should be considered as the first architectural response to keep a building cool. Shade is integral to the success of a courtyard. During the middle of the day when the Sun is highest, the extended eaves, covering external galleries and walkways, protect the interior floors and walls from direct solar radiation. Since the height of the courtyard is typically greater than it is wide, there is shading at the beginning and end of the day as the external walls shade the courtyard's exposed centre. Courtyards are often augmented with water (ponds or fountains) and vegetation to provide cooling from evapotranspiration. This can be further enhanced by spraying water on the courtyard floor a few times a day, which when it evaporates cools the surrounding air and building fabric; in dry climates this also raises the humidity to a more comfortable level. The Court of the Lions contains a central fountain, but also a series of channels to guide water throughout the courtyard and into the covered spaces at the edge (figure 3.14). This increases the evaporation of water within this enclosed space, cooling the air trapped in the courtyard, and thus enhancing the contained microclimate, but also cooling the high thermal mass floor coverings, thus reducing the radiant temperature.

Another important aspect of the Alhambra are the gardens and the extensive use of water throughout the landscape. Water and gardens have a special role in keeping surrounding buildings cool. Trees provide shade reducing temperatures of nearby surfaces (figure 3.12); hence, in larger courtyards, trees are used to help keep the environment cool and shade the courtyard floor, helping to reinforce the microclimate effects. However, water is perhaps the most important feature. To evaporate water

Figure 3.14. The Court of the Lions within the Alhambra, viewed from beneath the covered walkways, note the channels to guide water throughout the courtyard and additional fountains to wet the floor.

takes a lot of energy, and hence the evaporation of water is effective at cooling the local air and nearby surfaces. As stated above, water is a prominent feature in the courtyards at the Alhambra (and in courtyards elsewhere), but the Alhambra also shows other uses for water around buildings. The pool shown in figure 3.15 serves several purposes: it is decorative, it cools the air via evaporation, and it can also introduce light reflected off the surface into the rooms beyond. Also, in this warm dry climate, the evaporation of water increases the level of humidity to a more comfortable level.

The Alhambra, being a collection or royal palaces, gardens, and fortresses, is justifiably grand. The measures employed throughout the Alhambra are a result of the wealth of the occupants and the Mudéjar architectural style; these will not all be applicable to a dense urban setting or different locations. However, we can see variations on the responses showcased in the Alhambra across many other locations.

Across Mediterranean countries, the interior of buildings are kept cool during the day as they hold onto the cool of the night due to the high thermal mass of the walls and floors. At night or in early morning, the interiors are opened up and exposed to the cool air outside, purging any stored heat from the day before. Then, they are shut up again to keep direct light and warm air out during the day. This is especially important for upper floors, the bedrooms, which must be kept cooler than other living quarters in order to achieve thermal comfort suitable for sleep. Homes in

Figure 3.15. The Portico and nearby pool at the Alhambra (This Portico image has been obtained by the author from the Wikimedia website where it was made available under a CC BY-SA 4.0 licence. It is included within this article on that basis. It is attributed to Sebastian Appelt.).

many European countries employ mechanisms to keep the heat of the Sun out of interior spaces during the day, through use of shutters, curtains, blinds, awnings, or some other means such as plants (examples of these can be seen in figures 3.16–3.18).

The city of Valencia (Spain) offers further inspiration as to how to keep heat out of a modern city. The Old City of Valencia shows influences from the Middle East with compact stone buildings with few openings. However, as the city has expanded overtime, buildings have gotten larger with more windows for increased ventilation. Figure 3.17 shows two images of the more modern Valencia, with tall apartment blocks lining narrow streets. The height of these blocks is not purely a result of land prices; the height is matched to the street width so that the street below is perpetually in shade and buildings mutually shade each other. Despite this, all openings on the façades still have shutters, blinds, or awnings.

Another common sight within Valencia is the prevalence of roof terraces and gardens (figure 3.18). These outside spaces provide a useful additional living space for early morning or in the evening, when it is not too hot. Vegetation or awnings located on the terrace will shade the roof below and prevent heat from solar radiation entering the rooms below.

As we move around the Mediterranean we can see other examples of regional responses that have evolved to protect building occupants from the heat of the day. Santorini is the southernmost island of the Cyclades, a group of arid islands in the Aegean Sea between mainland Greece and the island of Crete. It is a popular tourist location largely due to its idyllic Mediterranean whitewashed architecture. The island of Santorini contains many examples of the Cycladic architectural style (figure 3.19). The many domed and vaulted roofs have a two-fold purpose: to shed water and heat effectively. Santorini receives little rainfall, which fall primarily during the winter months, and rainwater harvesting has been prevalent on the island

Figure 3.16. Houses in Cannes (France) with closed shutters to keep out the heat of the day.

Figure 3.17. Streets in Valencia (Spain), tall buildings frame narrow streets providing mutual shading.

Figure 3.18. Roof terraces in Valencia, providing an external living space for early morning and evenings.

for many centuries. All roofs are designed to shed water in some fashion. The vaulted roofs shed water very rapidly before it can evaporate, so that it can be stored in cisterns beneath the buildings. The curved roofs made of a high thermal mass material, such as concrete, also help keep the buildings cool. In hot sunny locations, intense solar radiation heats the building fabric, warming the interior spaces over the

Figure 3.19. Domed and vaulted roofs in Imerovigil on the island of Santorini (Greece) (This Imerovigil photo has been obtained by the author from the Wikimedia website where it was made available under a CC BY-SA 3.0 licence. It is included within this article on that basis. It is attributed to Berard Gagnon.).

course of the day. The curved surfaces help limit this warming by insuring that only a small area is directly facing the Sun at any given time, thereby reducing the intensity (W m^{-2}) of impinging solar radiation. External temperatures fall at the end of the day, and the vaulted and domes roofs have increased surface area, compared to a flat or pitched roof, over which to radiate heat away out to space. This increased radiative heat loss cools internal spaces in the evening, improving the thermal comfort of the occupants. It also purges the building fabric of heat accumulated during the day so that the cycle can begin again the next day.

So far we have considered the architectural responses found in warm dry climates and in warm maritime climates; if we go farther South there are further lessons we can learn from the tropics. The warm humid climate, potentially the most uncomfortable for humans, emphasises the need for shade, reduction of solar radiation on the east and west façades, but also the need to provide access for wind flow. This results in not only an elongated east–west form, but also a large roof overhangs and vegetation to provide shade.

Even quite light vegetative cover can have a beneficial effect in hot dry climate zones. Vegetation provides protection against glare, dust, and erosion, and assists in reducing air pollution. In tropical climates however, vegetation plays a further important role, acting as a wind break to protect buildings from high winds and driving rain. The use of vegetation as protection from the elements allows building façades to be very permeable with many openings (many without glazing). If we look at the islands that once formed British Malaya it is interesting to see how the local architecture has influenced traditional British architecture to produce a

Figure 3.20. Burkhill Hall (Singapore). Note the tall vegetation surrounding the building, the exaggerated roof overhang, and the awning to further protect the balcony and rooms beyond.

colonial style. Figure 3.20 shows Burkhill Hall located within the botanical gardens of Singapore. Burkhill Hall is considered the fore runner to the black and white colonial plantation houses that can be found across the region. Burkhill Hall sports a large roof overhang to not only shade the walls but to also protect them from driving rain. The roof coupled with the tall vegetation means the building is well protected from intense sunlight, heavy rainfall, and high winds. The building is surprisingly open, with a lattice work entrance, and in the photo we can see straight through the building to the gardens beyond. The upper floor does not even have a barrier between the interior rooms and the balcony.

Building materials are typically lightweight in this climate to allow the building to be cooled rapidly via ventilation. It is these features of sheltering and openness that characterises the buildings of the tropics; not just dwellings, but commercial and industrial buildings as well. Another more recent example can be found in a clothing factory in Sri Lanka. The MAS Intimates Thurulie facility claims to be the first clothing factory to be powered solely by carbon neutral energy sources, and has obtained one of the highest sustainability accolades: the LEED Platinum award. The factory was designed to harmonise with its site, with indoor and outdoor spaces integrated into a park. The site plan for the facility [3], shows rectangular factory buildings orientated with the east and west façades being the shortest. Meanwhile, communal areas are located on the north side of the complex, such as the cafeteria, which is open on one side (the north side) to the lake and park beyond. The lake and vegetation cool the air via evapotranspiration, which can then pass into the cafeteria and factory spaces, improving thermal comfort and worker productivity.

As part of its sustainability credentials, the factory was built from local soil, and was furnished and finished with indigenous materials such as bamboo. The factory complex incorporates traditional Sri Lankan architectural elements such as large roof overhangs for shade and protection from intense rainfall, courtyards, and covered walkways. In this way, there is a definitive link between the building, its occupants, and the surrounding landscape, which enhances not only the aesthetic but also the environmental conditions.

3.4 Summary

In this chapter, we have considered the effect of a warming climate on building occupants, considered what aspects of building design are available to help control internal conditions, and looked at what architectural responses have evolved in other climate zones. Consideration of how humans have learnt over time to live in different climates will be of great help when considering how to adapt new and existing buildings to a changing climate. A summary of regional responses can be found below:

Warm Maritime—e.g. Mediterranean	
Climate	Summers are warm to hot with little to no rain, winters are cool to cold with moderate rainfall. Intense solar radiation, especially in summer. Variation in temperatures, large in continental locations, small in coastal/marine locations.
Layout and form	In continental locations the response is similar to hot dry zones, with compact buildings and urban areas for mutual shading, while more marine climates buildings are spread out more to allow cooling sea breezes to permeate between buildings and convect heat away. Buildings preferably one room deep in marine areas to allow for good air flow. Courtyards used for same effect in continental areas and also provide protection from cold winter winds.
Structure	For continental locations see hot dry zones. For maritime climates, thermal mass is not as important since the sea provides thermal regulation; however, floors and roofs are typically heavyweight to store heat from the Sun in winter. Roofs are designed to shed intense rainfall. All surfaces are light in colour to reflect solar radiation.
Hot dry—e.g. Middle East	
Climate	Lots of intense solar radiation, leading to large ranges of diurnal and annual temperatures. Low humidity and low precipitation. Need to consider glare off of surfaces.
Layout and form	Buildings are compact and located close to each other for mutual shading. Buildings are enclosed and inward looking with courtyards and few external openings. Larger buildings are cuboid shapes to minimise surface area, to limit solar receipt. Spaces near east and west façades are non-living spaces to provide a thermal barrier against the low sun angles. Rooms are deep, and covered walkways used to prevent heat being reflected or radiated in to rooms. Fountains or water features used to cool patios and courtyards. Flat roofs or domed roofs used to limit solar receipt and maximise radiative heat loss.

Structure	Windows are relatively small, particularly on external walls, and are shaded from direct solar radiation by shutters or shades. Walls are typically thick and made of thermally massive materials. The roof is solid with a reflective upper surface, and sometimes a ventilated double roof is employed with the upper roof shading the roof below in contact with interior spaces. Surfaces are light in colour, sometimes with white wash to maximise reflectivity.
Warm humid—e.g. Sri Lanka	
Climate	Lots of rainfall with high humidity. Smaller diurnal and annual range of temperatures with a high even temperature all year round. Generally, only low winds speeds. Intense solar radiation due to latitude, with a high diffuse proportion due to the humidity, leading to high sky glare. Intense rainfall and often electrical storms in the afternoon.
Layout and form	Buildings are elongated in the east–west axis to minimise low solar angle sunlight. Urban layouts are sparse to allow what little wind there is to penetrate between buildings. Buildings are preferably only one room deep, opening onto verandas, galleries, or balconies to allow for better ventilation penetration. Buildings occasionally built on stilts to access greater wind speeds away from the ground. Living areas are typically open plan to allow for better ventilation. Externally large roofs are pitched to efficiently shed rainfall with large overhangs to protect against sky glare and protect the building façades from intense rainfall and sunlight. Tall vegetation is also used to shelter the building.
Structure	Openings are large to allow for adequate ventilation. Sections of façade are openable either by folding walls, or louvered openings (Jalousie windows); other sections of façade may be permeable screens or latticework to provide privacy but still allow airflow. Materials are typically lightweight with low thermal mass to allow rapid cooling from airflow. Double skin ventilated roofs often used to limit solar receipt through the large roof. Buildings are generally light in colour to increase reflectivity, but the humid climate can lead to algae and fungi growth, thus limiting the effectiveness of this measure.

References

[1] Roelofsen P 2002 The impact of office environments on employee performance: the design of the workplace as a strategy for productivity enhancement *J. Fac. Manag.* **1** 247–64

[2] Liu C, Kershaw T, Fosas D, Ramallo Gonzalez A P, Natarajan S and Coley D A 2017 High resolution mapping of overheating and mortality risk *Build. Environ.* **122** 1–14

[3] MAS Intimates Thurulie Facility https://www.lafargeholcim-foundation.org/Publications/mas-intimates-thurulie-clothing-factory-in-sri-lanka

Chapter 4

The urban heat island (UHI)

For the first time in history, more than half the world's population lives in urban areas. This is a recent phenomenon; in 1900 only 14% lived in urban areas, but this number is expected to grow to 60% by 2030 and 66% by 2050. Despite this rapid urbanisation, cities and urban areas occupy less than 2% of the Earth's land surface. To accommodate this rapid urbanisation, we will be seeing the creation of new cities and substantial growth of existing cities over the next few decades. The greatest rates of urbanisation are anticipated to be in Africa and Asia, where some countries are expected to experience greater than a five-fold increase in urban populations by 2050. These new and expanding cities will face a series of challenges from a changing climate, such as changes in rainfall, temperature, increases in the frequency and severity of extreme weather events, and potentially shifting weather patterns. The situation is made more complex as urban areas also exhibit their own microclimate arising from the presence of buildings and human activity.

It has long been known that cities generate their own microclimate and are typically warmer than their rural surroundings. This 'mesoscale' effect is known as the urban heat island (UHI). Luke Howard (who is best known for his cloud classification system) first showed the existence of this urban microclimate and measured the temperature difference between London and its surroundings to be 2 °C 200 years ago. The UHI is not constant; it varies according to weather conditions, season, and the time of day. Additionally, the UHI is not constant across a whole urban area, and varies according to the size and proximity of buildings. The heat island effect is generally greatest during calm, warm weather, and is most pronounced at night. Since Luke Howard measured the UHI of London in 1818, the city has grown. By the 1960s, the UHI had increased to a ~5 °C difference between the city centre and the rural surroundings, and by the end of the last century the UHI had increased to ~7 °C. This temperature difference arises due to a number of physical processes, which are the subject of this chapter.

doi:10.1088/978-0-7503-1197-7ch4

As you might expect, the UHI has a profound effect on the way buildings within cities behave, and hence on the comfort and health of building occupants. The origins of the heat island lie in the fact that buildings, roads, and pavements store heat gained during the day, both from solar radiation and from human-related activity such as traffic exhaust and heat loss from buildings. The combination of the city geometry, which traps heat at street level, and reduced hydrodynamical cooling from hard, impermeable surfaces, are coupled with improved drainage to quickly carry water away before it can evaporate. Additionally, cities tend to have fewer trees and green areas compared to rural areas, resulting in less cooling from the evapotranspiration of water from plant leaves and the soil.

Urban areas have a larger surface area than rural areas over which to absorb solar radiation, but also to lose heat to the atmosphere by radiation, convection, conduction, and evaporation. However, the geometry of cities leads to the formation of atmospheric boundary layers, which limit the vertical transport of heat away from the cityscape. The UHI represents a partitioning of the atmosphere above urban areas exhibiting reduced vertical transport of air, caused by changes in surface roughness, temperature and humidity gradients. The UHI can be considered to be made up of three principle components:

- Urban surface layer: the heated surfaces of buildings, roads, and other surfaces in the urban area.
- Urban canopy layer (UCL): extends from the surface to the rooftops or the tops of tree canopies.
- Urban boundary layer (UBL): extends upwards from the top of the canopy layer as a result of heat and moisture transfer between the urban area and the air above. It acts as a mixing layer between the urban environment and the surrounding atmosphere, and can be likened to a bubble of warm air sitting on top of the urban area.

4.1 Boundary layer creation

The Earth's atmosphere is defined by many different layers. The lowest level of the atmosphere, the homosphere, which extends up to ~100 km, represents the part of the atmosphere where atmospheric gases are considered fully mixed. Within the homosphere there are four sub-regions: in descending order, the thermosphere, mesosphere, stratosphere, and the troposphere. The troposphere extends upwards from the surface to ~10 km, and contains ~75% of the Earth's atmospheric mass and the vast majority of the planet's water vapour. This layer is further partitioned into the free troposphere and the planetary boundary layer (PBL). The PBL is the part of the troposphere that is influenced by contact with the planetary surface. The height of the PBL depends on the strength of the atmospheric mixing generated by the surface. Typically, its depth varies according to a diurnal (24 h) and seasonal cycle. During the day, solar radiation generates strong thermal atmospheric mixing as a result of convection currents that extend the PBL up to ~1–2 km, while at night the cooling of the surface relative to the atmosphere causes a downward flux of heat, supressing atmospheric mixing, and the PBL contracts to <100 m. The same thermal processes contribute to a greater mean PBL depth during the summer compared to winter.

Generally, a boundary layer can be considered as a layer in a fluid (i.e. a gas or liquid), where transfer processes are influenced by the properties of the underlying surface. A discussion of these processes and momentum transfer is therefore needed as background to this chapter in order to consider different aspects of exchange between urban areas and the atmosphere above.

Figure 4.1 shows the formation of a boundary layer (red line) over a smooth flat surface immersed in a non-turbulent moving fluid. When the direction of flow is approximately parallel to the surface, the flow is said to be *laminar*. In laminar flow, momentum transfer is between individual molecules. As the thickness of a laminar boundary layer increases, it becomes unstable and eventually breaks up to forming chaotic swirling motions, termed the *turbulent boundary layer*. A second laminar layer of restricted depth, termed the *laminar sublayer*, forms immediately above the surface and below the turbulent layer. The transition from laminar flow to turbulent flow depends on the ratio between the inertial forces associated with the flow over the surface and the viscous forces generated by the inter-molecular attraction, sometimes called internal-friction. The ratio between the inertial and viscous forces is called the Reynolds Number (Re), after the British scientist Osborne Reynolds who first showed that the onset of turbulence for liquid in a pipe depended upon this ratio. When Re is small, viscous forces dominate and the flow tends to remain laminar, but when the ratio increases beyond a critical value (Re_{crit}), inertial forces dominate and the flow becomes turbulent.

The Reynolds number can be calculated from $Re = Ud/v$, where U is the free-stream velocity, d is determined from the dimensions of the system, and v is the

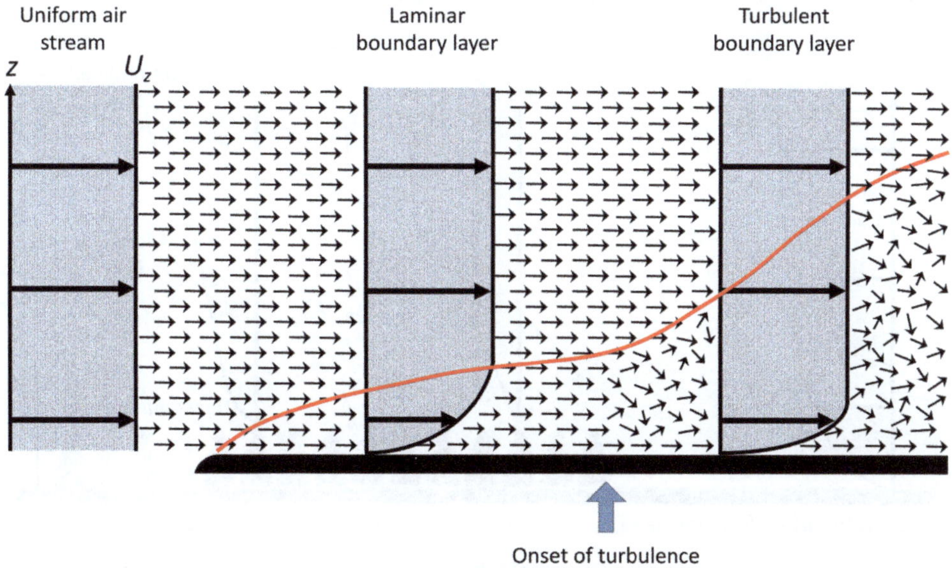

Figure 4.1. Formation of laminar and turbulent boundary layers, and associated velocity profiles for a fluid flowing over a smooth plate (the vertical scale is highly exaggerated).

coefficient of kinematic viscosity, a property of the fluid. As a fluid flows over a flat plate, as in figure 4.1, d is the distance from the leading edge. For a very smooth plate exposed to a non-turbulent, parallel flow of air Re_{crit} is of the order of 10^6.

In order to conserve momentum across the boundary between the fluid and the plate, the fluid velocity has to be zero at the surface. For both in laminar and turbulent boundary layers, the velocity increases from zero to the free-stream value U at the top of the boundary layer. The velocity profiles shown in figure 4.1 indicate that in a laminar boundary layer the velocity increases constantly to U at the top of the boundary layer, while for the turbulent boundary layer, the velocity reaches U at the top of the laminar sublayer. Remember that velocity is a vector and not a scalar value, so direction is important. In terms of figure 4.1, the maximum velocity in direction z is achieved by the top of the laminar sublayer and above in the turbulent boundary layer this component of velocity (U_z) is maintained, but since the fluid flow is turbulent there are other velocity components as well.

For a fluid in laminar flow over a relatively smooth surface, movement is predictable and the resultant boundary layer has well-defined profiles of velocity, vertical flux, and temperature. In contrast, turbulent flow is unpredictable both spatially and temporally. We have seen in figure 4.1 how the depth of a laminar boundary layer above a flat plate increases with distance from the leading edge. In the same way, the depth of a turbulent boundary layer depends on the 'fetch' across a uniformly rough surface. Figure 4.2 illustrates how the wind profile varies as flow moves from a relatively smooth surface such as open grassland to a rougher surface such as forest. The rougher surface exerts more drag than the grass, and the air is decelerated as it moves to the rougher surface. The boundary layer that had formed over the smooth surface is disrupted by the change in roughness, creating a layer of modified flow. As the flow progresses over the rougher surface, a new equilibrium

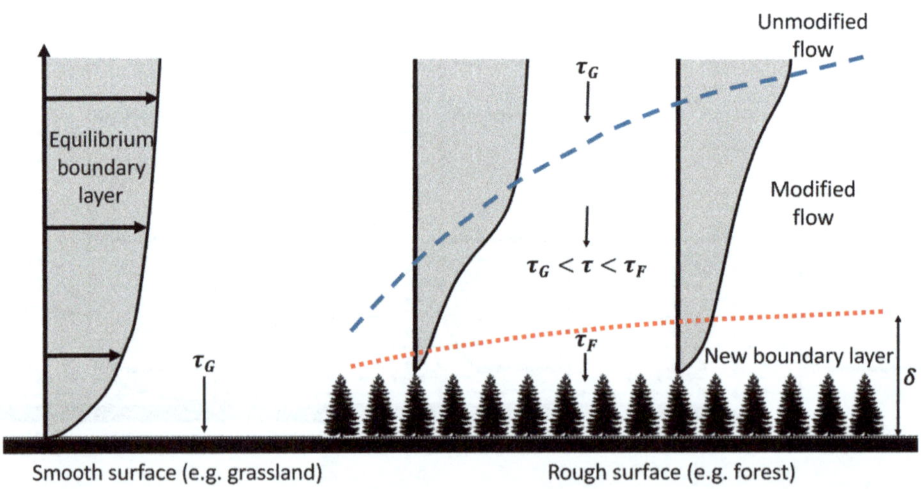

Figure 4.2. Development of a new boundary layer (dotted line) as air flows from a smooth surface to a rougher surface. The dashed line represents the boundary between the modified and unmodified flows where the vertical momentum flux is τ_G (not to scale).

boundary layer develops with depth δ, in which velocities, vertical flux (τ), and temperatures are characteristic of the rougher surface. The depth δ can be linked to the upwind fetch x. Between this new boundary layer and the unmodified flow above is a mixing region where flows are intermediate between those of the two different surfaces.

The depth of a boundary layer δ has been measured many times, and while estimates vary, the ratio between the fetch x and δ are in the range of 20:1 to 100:1. Within this new boundary layer, there are two sublayers. Very close to the rough surface (in this case a forest) this homogeneity is disrupted by the wakes of individual roughness elements (trees), leading to the formation of a *roughness sublayer*, whose properties are determined by the structure of roughness elements (e.g. foliage and plant spacing). Above this roughness sublayer is the *inertial sublayer*, within which fluxes (τ) of heat, water vapour, and other gases are constant with height, provided that the flow is fully turbulent. In the inertial sublayer, which forms at ~2× the canopy height, the only variables of importance are the friction velocity and height. This constant flux layer constitutes only the bottom ~15% of the atmosphere, which is influenced by the nature of the surface (the PBL).

While the examples provided above relate to boundary layers forming above natural surfaces, it is easy to imagine how similar effects can be observed above cities and urban areas. This means that the geometry of our buildings and streets, which constitute the surface roughness of an urban area and hence influence friction velocities, the depth of sublayers, and vertical flux have a profound impact on the environment experienced within our cities. Boundary layers isolate the cityscape from the atmosphere above, limiting the convection of heat and pollutants away from the city.

4.2 The energetic basis and UHI creation

The increased surface roughness of a cityscape compared to its rural surroundings leads to the formation of a new boundary layer, partitioning the PBL into the UBL and the UCL. The UBL is a mesoscale effect, ranging from the unmodified PBL above to the roofs of the cityscape below. The properties of the UBL (vertical flux, wind velocities, and temperature gradients) are influenced by the presence of an urban area at its lower boundary. The UCL, however, is a microscale effect that describes the lowest part of the PBL comprising of the air between urban roughness elements (buildings, roads, trees, etc). The environmental conditions experienced within the UCL are dominated by the materials and morphology of the immediate surroundings and human activities. The UCL represents the anthroposphere in urban areas; it is the part of a city that humans inhabit, and is consequently the part of the atmosphere that is vital for ensuring human health, wellbeing, and comfort in cities.

This simplified representation of the urban climate that includes the UCL with the UBL above is blurred by the presence of air turbulence caused by wind drag and shear stresses as air flows over the buildings. As with the forest example used previously, mechanical turbulence is induced by the built-environment's surface roughness, producing groups of eddies of varying scales. This creates a *roughness sublayer* that extends up to about twice the height of the buildings and includes the UCL. Above this roughness sublayer is the *inertial sublayer*, which extends up to the top of the *surface layer*. The reminder of the UBL is comprised of a mixing layer where environmental conditions vary from those of the urban surface at the lower bound to that of the unmodified rural surface at the top (see figures 4.2 and 4.3).

As with the global climate, the primary driver for the urban microclimate is the Sun. Urban areas differ from rural areas in that they have a lower albedo (are less reflective to solar radiation) and are constructed from high thermal mass materials. This means that considerably more solar radiation can be absorbed and stored in an urban area than in a rural one. During the day, solar radiation warms the surface (land and buildings) and the surrounding air, inducing convective instability resulting in the vertical flux of air parcels or thermals rising up into the UBL. As these thermals rise, they mix with the atmosphere above to form a boundary layer with constant temperature and almost neutral stability. This buoyancy-driven process drives warmer air to the top of the boundary layer to form a stable capping inversion layer (warmer fluid above a relatively cooler fluid). This temperature inversion and stable capping layer prevents thermals rising from below, trapping heat, water vapour, and pollutants released at the surface within the boundary layer.

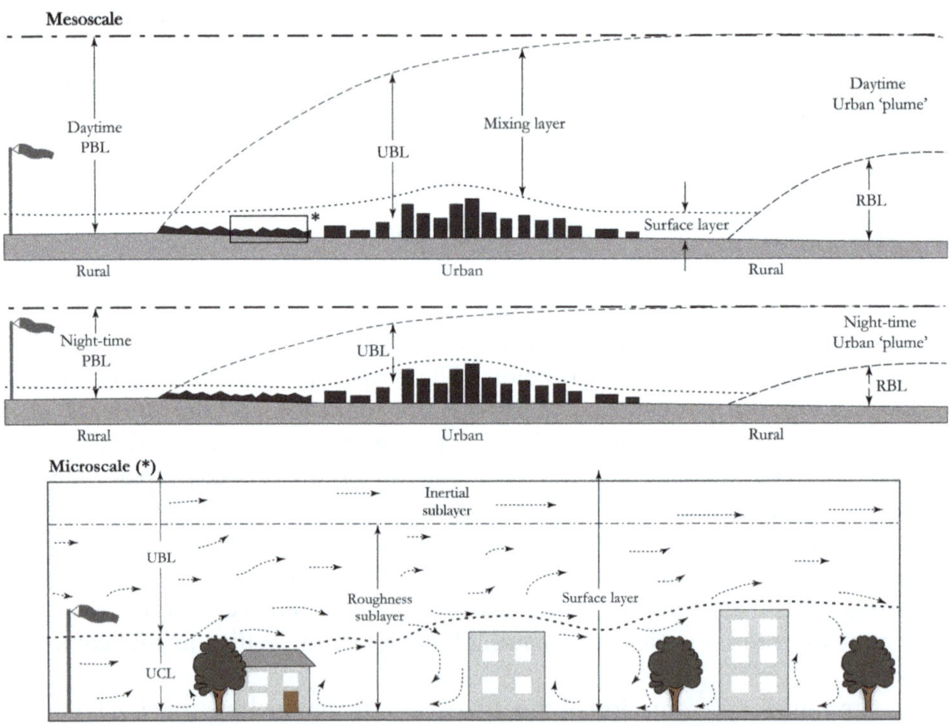

Figure 4.3. Boundary layer structure over a cityscape arising from changes in surface roughness. Top, daytime boundary layers; middle, nocturnal boundary layers; bottom, close up of the microscale structures showing approximate locations of the roughness and inertial sublayers that make up the surface layer.

The primary difference between urban and rural boundary layers (RBL) is characterised by the intensity and depth of this capping inversion layer. The lower albedo of the cityscape compared to rural areas results in the release of more heat during the day, causing the stable capping inversion layer to be warmer and thicker than in rural areas. This boundary layer heat island is the result of the relative difference between rural and urban warming.

During the night, the surface loses the heat it had gained from solar radiation during the day to the atmosphere above. Once the surface has cooled, there is a downward flux of heat from the warm atmosphere above, causing the boundary layer to contract. Rural landscapes are quick to warm when exposed to solar radiation, but are also able to purge accumulated heat quickly at night via evaporation, radiation, and convection. Urban landscapes, however, are slow to warm during the day due to mutual shading from buildings and the high thermal mass of the materials present. However, once warmed they are slow to lose accumulated heat to the atmosphere. The geometry of the streets traps heat at street level. Solar radiation comprises of both direct and diffuse radiation. Diffuse radiation arises from scattering in the atmosphere and provides equal intensity on all surfaces regardless of orientation, while direct radiation depends upon the position of the Sun. However, when heat is radiated from a surface it is entirely diffuse, being emitted in all directions equally (figure 4.4). As such, depending on the height and width of streets, heat radiated from a warm building has only a small chance of

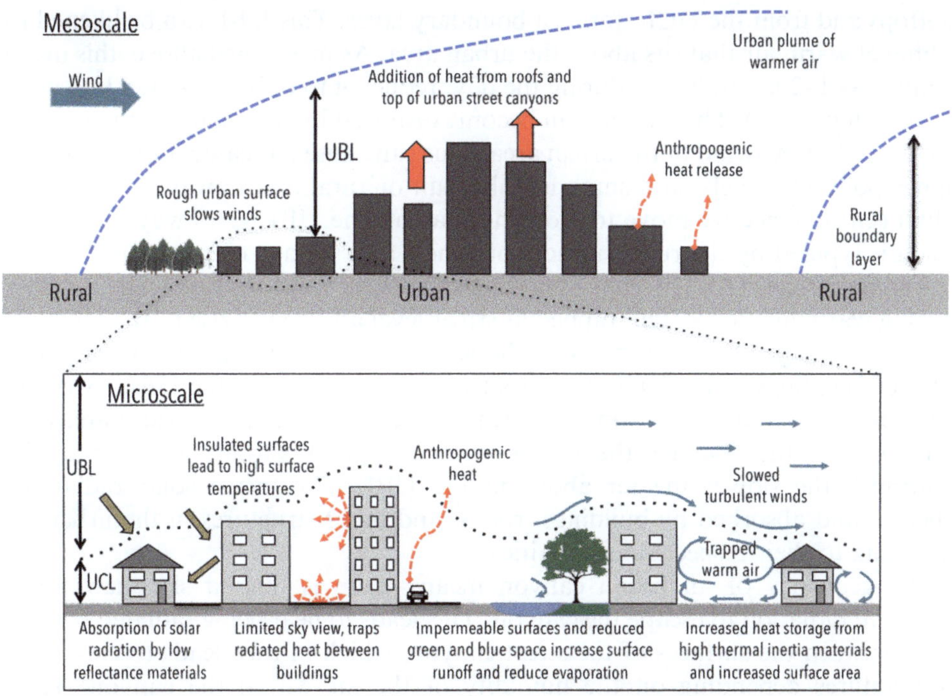

Figure 4.4. Illustration of the energetic processes that lead to the UHI, showing the UBL (top) and the UCL (bottom).

being radiated out to space. Instead, it may be radiated towards the ground, a road, or another building. Therefore, heat may be radiated and reabsorbed many times before it can escape from street level, an effect termed the canyon effect.

Due to this canyon effect and the high thermal mass of urban areas, urban areas continue to emit heat well into the night and continue to generate thermals, although at a lesser extent than during the day and decreasing in altitude, thus preventing the breakup of the capping inversion layer. The persistence of this capping layer, which limits heat loss from the cityscape below, means that the greatest temperature differences between rural and urban areas are often experienced during the night when rural temperatures are at their minimum.

The UHI that is experienced in a real city is not only a product of the formation of a boundary layer arising from changes in surface roughness and the trapping of heat due to the formation of a stable inversion layer. The addition of anthropogenic heat (from car exhausts, building heating, ventilation, and cooling systems and even metabolic heat) further exacerbate temperatures. Coupled with a reduction in evaporation from improved drainage and reduced trees and ponds, which also reduces the vertical flux of air, heat, and other gases (water vapour is less dense than air and rises naturally) means that the urban microclimate can be significantly different from the rural climate. A summary of the energetic processes that go into the formation of the UHI can be seen in figure 4.4.

To summarise, at a mesoscale (10^3–10^5 m) or citywide scale, the rough urban surface slows wind speeds and with the addition of anthropogenic heat, from rooftops and from the UCL, forms a boundary layer. This UBL can be likened to a bubble of warm air that sits above the urban area. As mentioned above, this may be as much as 1–2 km in height during the day, reducing to only a few hundred metres at night (figure 4.3). This bubble can become distorted by the wind, forming a plume that extends downwind of the urban area. This can cause increased air temperatures, higher pollution levels, and smog in suburban or rural areas away from the main urban centre. Once we move to more rural areas, the UBL gives way to the RBL, which is typified by the rural surface roughness and surface temperatures.

At a microscale (10^2–10^3 m) or building/street scale, airflow is typically turbulent, causing warm air to become trapped at street level and giving rise to the UCL. The greater the height to width ratio of the street canyons, the more isolated the air between buildings in the UCL becomes from the air above in the UBL. Airflow in the UBL is generally more laminar albeit at a reduced speed than rural areas for a given height; the rougher the urban surface (greater surface area for a given footprint), the slower the air above in the UBL. Shortwave solar radiation is reflected and absorbed by buildings, roads, and other structures in the urban area according to their respective emissivities.

The geometry of an urban canyon means there is limited sky view for the longwave radiation to escape the canyon. This leads to heat being radiated back and forth between the surfaces of the street canyon, which in turn leads to elevated air temperatures depending on the humidity of the air within the canyon. This is exacerbated by anthropogenic heat from car exhausts, air conditioning systems, flues, etc. The surface roughness and the level of isolation between the UCL and the

UBL, as well as rural wind speed, determine how quickly warm and polluted air in the UCL can dissipate.

4.3 Weather influence

As you might expect, the weather has a profound influence on the magnitude of any UHI experienced. Weather patterns can significantly modify the heat transfer processes between the urban surfaces and the atmosphere, with wind velocity followed by cloud cover being the main variables to consider.

As mentioned in chapter 1, wind is the product of temperature variations across the face of the Earth, leading to large-scale global convection currents, which generate pressure differences resulting in atmospheric motion. At the macroscale (continental level), wind speed and direction is influenced by the presence of land masses and topography that can deflect weather systems and determine wind speeds at different levels. At a mesoscale, wind speeds are influenced by the surface roughness of the urban form, which alters flow patterns and wind velocities with altitude. Meanwhile, at a microscale, variations in surface temperatures and wind sheltering from buildings are the dominant features. Wind velocity is the most important variable for reducing the UHI intensity; higher wind speeds (particularly over a rough surface such as a city) increases turbulence and hence advective activity. This has the effect of disrupting the temperature inversions that intensify the UHI, resulting in a more homogenous vertical temperature profile.

After wind, cloud cover is the most important weather variable that affects the UHI, as it influences how much solar radiation reaches the surface. Incoming shortwave (visible) solar radiation is absorbed and reflected by cloud tops, while heat radiated from the surface is absorbed and re-emitted by the clouds, which due to their high water content act as almost perfect black bodies in the thermal infrared part of the spectrum. Hence, increased cloud cover reduces both incident radiation (heating) and surface heat loss (cooling). The amount and type of cloud cover is important. Lower altitude, thicker stratus type clouds have a greater influence than higher altitude, thinner cirrus type clouds. Thick cloud cover can reduce incident solar radiation to 10% of the that on a clear day. Hence, cloud cover can have a great influence on the surface temperatures experienced in an urban area. The overall effect of cloud cover is to moderate the variation in the surface radiation balance, resulting in a smoother, flatter diurnal temperature range.

Both wind and cloud cover can be considered as representative of atmospheric stability, a measure of the vertical transport of air (see figure 4.5). High wind velocities and cloud cover are associated with increased atmospheric stability and reduced vertical transport. High wind velocities achieve this through increased advection and turbulence (termed dynamic instability), while cloud cover reduces the scope for convective thermals to carry heat and air into the high atmosphere. In contrast, calmer weather with low wind speeds and clear skies, leads to a PBL that is generally considered unstable during the day and stable at night. In reality, overall atmospheric stability must be considered as being made up of different layers of varying stability. As discussed earlier in this chapter, the formation of the UHI

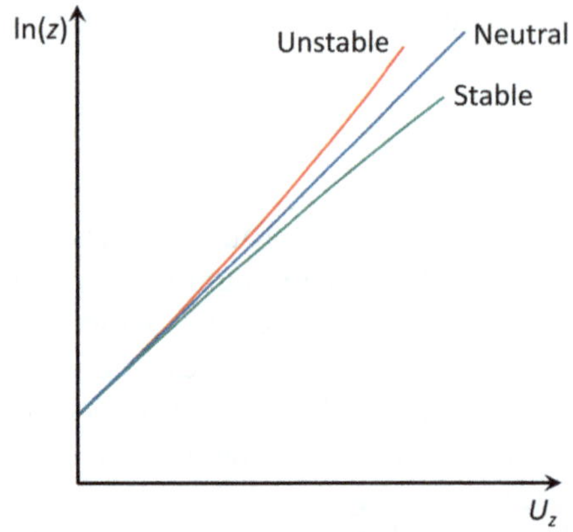

Figure 4.5. Vertical wind speed (U_z) as a function of height (z) for the three basic atmospheric stability states for airflow near the ground.

results from convective thermals rising from an unstable surface layer, followed by a neutral mixing layer and a stable capping thermal inversion layer.

Dense urban forms act as a windbreak decreasing mean wind velocities. This reduction occurs from the transformation of the wind's mean kinetic energy into turbulent kinetic energy. Greater surface roughness from the built form and vegetation generates greater transformation of flow; this increases turbulent diffusivity, enhancing both sensible and latent flux regardless of the vertical temperature and water vapour gradients. Strong winds promote dynamic instability and turbulent mixing, inhibiting the development of strong temperature stratification, while reduced wind velocities have the opposite effect and less heat is transferred from the surface to the atmosphere. Forced convection generated by surface roughness decreases more rapidly with altitude than the effects of thermal turbulence. These mechanical turbulence effects are therefore primarily dominant near the surface, while at greater altitudes, thermal turbulence effects dominate. This is reflected in the diurnal wind profiles of the atmosphere. During the day, most of the upper boundary layer is dominated by free convection with large eddy currents resulting from thermals. At night however, the increased stability of the atmosphere and absence of thermals means that convective mixing in this shallower boundary layer is entirely mechanical in nature with relatively smaller eddy currents.

High pressure anticyclonic weather systems affect temperatures in the PBL by causing air to subside and warm on top of a cooler surface air to create a subsidence inversion layer. This descending warmer air causes clouds to evaporate, thereby increasing shortwave solar radiation penetration through the atmosphere and increasing surface temperatures, and leading to atmospheric instability and vertical transport ideal for UHI formation. Conversely, low-pressure systems cause air to rise and cool, generating clouds and reducing shortwave solar radiation penetration.

The most intense UHIs have been observed during persisting high pressure weather systems. Unfortunately, such weather systems that bring clear skies and intense solar radiation are also typical of high air temperatures. The UHI of London was measured to be as high as 9 °C during the August 2003 heatwave (compared to ~7 °C normally on clear summer days) and was likely instrumental in the increased heat related mortality rate observed during this period. As the climate changes and we see warmer drier summers, the formation of high-pressure anticyclonic weather systems is likely to increase and will have an influence on future observed UHIs.

4.4 Implications of the UHI on the built environment

The UHI alters the daily cycle of temperatures in an urban environment, both in magnitude but also shifted in time. Figure 4.6 shows the difference between recorded temperatures at a rural location and a those simulated for a dense urban location, in this case the city of London[1], for a clear summers day with an intense UHI. As we can see, the effects of the urban environment lead to elevated nocturnal temperatures. However, during the day urban temperatures can be lower than in rural locations as a result of the mutual shading of buildings and the high thermal mass of the materials present in urban locations; this is termed as the urban cool island effect. We can also see that daytime peak temperatures are similar. This is likely due to the fact that the stable capping thermal inversion layer above the city has yet to

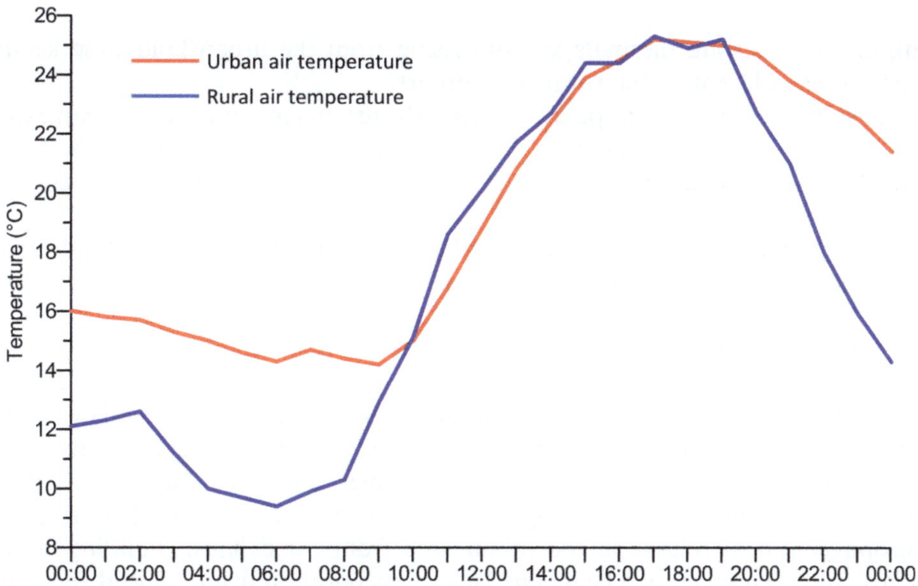

Figure 4.6. Variation between urban and rural temperatures on a clear summer's day.

[1] Data presented in this section compiled from simulations detailed in [1].

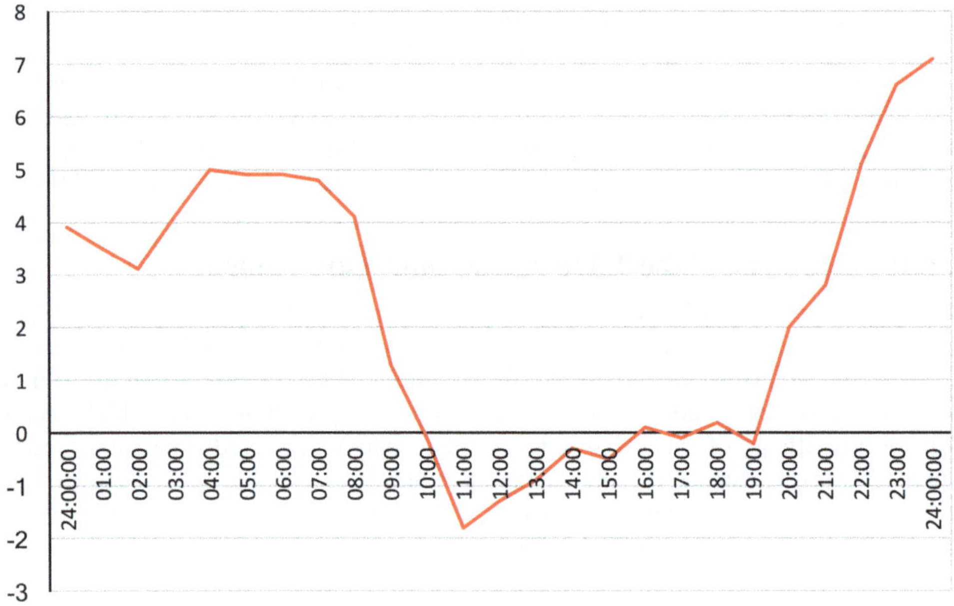

Figure 4.7. The diurnal UHI cycle, showing the difference between the urban and rural temperatures shown in figure 4.6.

form, and thus strong thermals are still rising from the urban landscape keeping temperatures in line with the rural counterpart.

If we subtract the rural temperatures from the urban temperatures, we can extract the effect of the urban environment (figure 4.7). We can see that at night the UHI is as large as 7 °C while during the day the urban area can be as much as 2 °C cooler in the late morning.

The magnitude of the urban heat and cool islands depends on latitude (solar angle), the size of the buildings, anthropogenic heat generation in the urban area, and the different factors that govern the formation of the boundary layers. The values presented in figure 4.7 are in line with those measured for the city of London. As mentioned previously, the weather also has a profound impact upon heat island generation. Figure 4.8 shows recorded rural temperatures and simulated urban temperatures for a completely overcast summer's day. Here we can see no distinct heat island or cool island; instead, the urban temperatures are systematically higher than the rural temperatures throughout the whole diurnal cycle. This is due to anthropogenic heat release in the urban area from car exhaust, building cooling systems, and even metabolic heat. The lack of evaporation from reduced vegetation and waterbodies in the urban area also limit vertical transport of heat, resulting in slightly elevated temperatures.

The UHI in cooler climates, such as that of the UK currently, can be beneficial since buildings in cities require less winter heating energy than their rural counterparts. Additionally, the cool island during the morning can move summertime peak

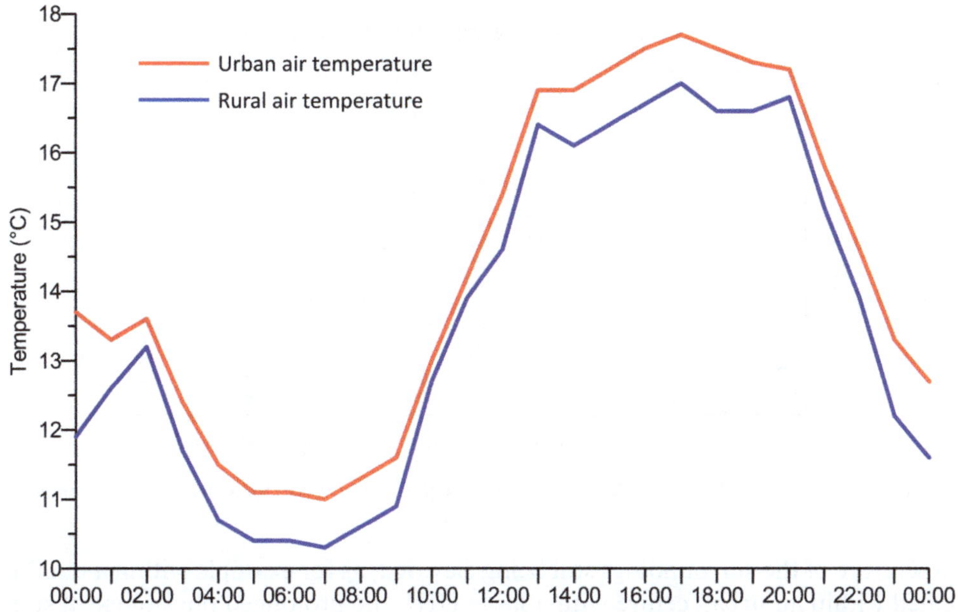

Figure 4.8. Variation between urban and rural temperatures on a completely overcast summer's day.

internal temperatures towards the end of the day and potentially outside of normal working hours. However, this means that the UHI decreases the ability of buildings to shed heat at night, significantly reducing the effectiveness of night-time cooling. This will have an impact on thermal comfort in domestic properties, which are occupied at night.

Continuing with our city of London example, we find that the presence of the UHI has an effect on the amount of energy used for heating and cooling throughout the year. Figure 4.9 shows the heating and cooling energy for a medium-sized commercial office exposed to a rural climate and an urban climate. While savings are made in winter with respect to reduced heating load, this is repaid in summer through increased cooling loads. Whether the UHI has a net positive or negative effect on the energy performance of the offices depends upon the coefficient of performance (CoP) of the heating and cooling systems employed.

In addition to the energy implications of the UHI, there are also physiological implications. During the 2003 European heatwave, which claimed up to 70 000 lives, the highest temperatures were not recorded in urban areas; however, this is where the people died. The UHI in London was measured to be as much as 9 °C during this heatwave due to the clear sunny weather. As we saw from figures 4.6 and 4.7 the heatwave is a maximum at night, and herein lies the danger. This is because human physiology allows us to be exposed to high temperatures for short periods of time (hours). However, over longer time periods, the physiological strain of increased sweating, blood vessel dilation, and increased core body temperature can have serious health implications. If the body is not allowed to cool overnight due to the

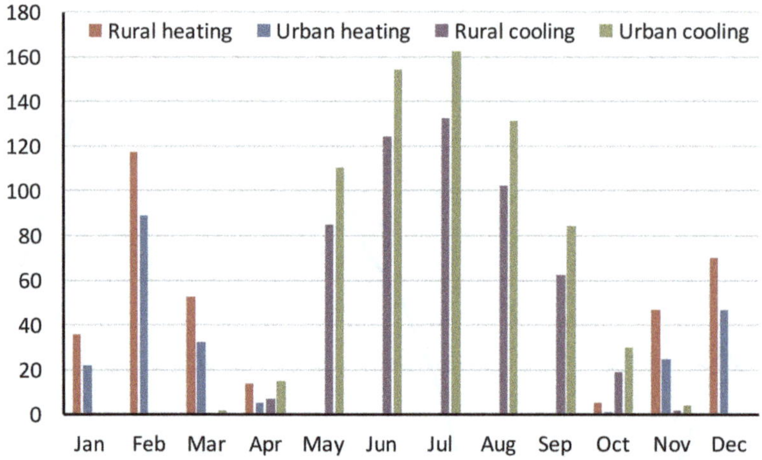

Figure 4.9. Implications of the UHI on both heating and cooling loads (MWh) for a medium-sized commercial office.

temperature of the surroundings still being elevated, as for example within a building located within an urban centre with a large UHI, the prolonged physiological stress can cause not only severe dehydration but also a variety of heat-related illnesses including heart failure.

4.5 Air quality in cities

The UHI is not just a threat to human life during hot weather and heatwaves, but is also is associated with poor air quality. In the UK, around 40 000 deaths per year are attributable to outdoor air pollution, and the World Health Organisation (WHO) estimate that globally over seven million deaths per year are a result of air pollution. Air pollution plays a role in many of the major health challenges facing the global population. Air pollution has been linked to various cancers, asthma, heart disease, and strokes, as well as diabetes, mental illnesses such as dementia, and increased obesity.

The health implications of air pollution accrue over a lifetime, from a baby's first few weeks in the womb, through adulthood to old age. Exposure to air pollution during pregnancy has been associated with low birth weights and premature births. Gestation, infancy, and early childhood are particularly vulnerable times because the young body is developing rapidly and the heart, brain, immunological, and hormone systems can all be harmed by air pollution.

Environmental variables such as temperature, humidity, precipitation, wind speed/direction, air pressure, and the mixing height (the top of the surface layer, see figure 4.3), all have important roles in determining the air quality above a city. When a high-pressure weather system sits above an area, it concentrates pollutants close to the ground, suppressing vertical diffusion. High levels of UV radiation during warm sunny weather reacts with pollutants to produce ground-level ozone (O_3). O_3 can cause shortness of breath, coughing, wheezing, decreased lung capacity, inflammation of the airways, and even death. Anthropogenic emissions such as oxides of nitrogen (NO_x) and volatile organic compounds (VOCs) contribute to O_3 production. In addition to these health impacts, some gases act as greenhouse gases, while others such as sulphur dioxide (SO_2) have a cooling effect on the local environment. Gaseous pollutants can be dispersed via horizontal wind flow and via vertical transport of air up from the surface. Rain is necessary to clear pollutants out of the air, and without rain, the concentrations of dust, smoke, and other pollutants (particularly particulate pollutants) will increase over time. This presents an important side effect of the shifting of rainfall from the summer to the winter as a result of climate change. As the climate warms and summers become drier, we can expect urban air quality to worsen, and exacerbation of the associated negative health impacts.

In Luke Howard's day, air pollution could largely be attributed to the burning of coal, and the dark smoke it produced that hung over cities was highly visible. However, modern-day air pollution in the UK largely comprises of very small particulate matter, oxides of nitrogen, and O_3. These are odourless and invisible to the naked eye except for an occasional brown haze from the oxides of nitrogen that can be seen hanging over cities. Modern-day pollutants have been shown to be highly damaging to not only the lungs but also the heart and cardiovascular system. It is partly for these reasons that many cities have introduced congestion charges, and several European cities are considering banning diesel vehicles by 2025 due to their production of particulate matter pollution. The above images are from the

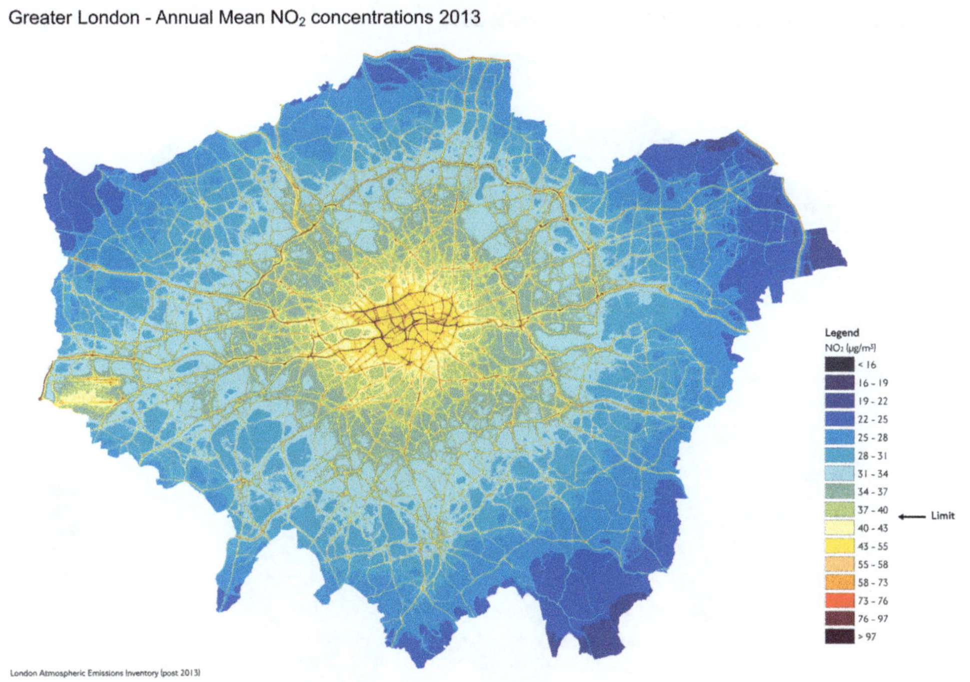

Figure 4.10. Annual mean nitrogen dioxide (NO_2) concentration. Note how much of London exceeds the national limit for NO_2 (courtesy of LAEI).

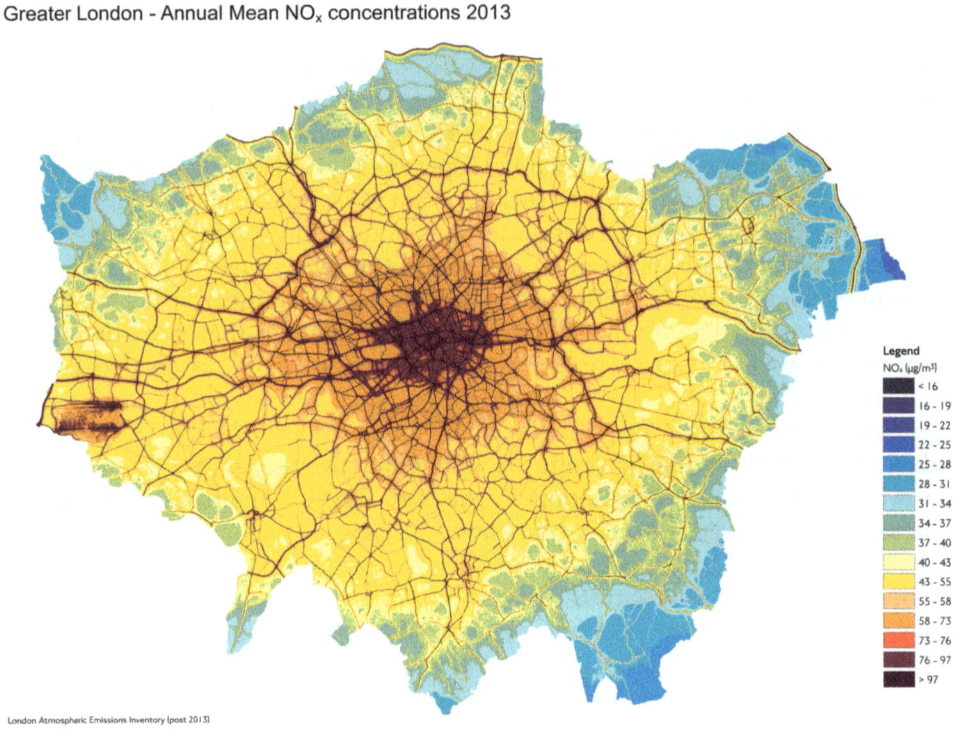

Figure 4.11. Annual mean concentration for the ensemble of Nitrogen oxides (NO_x) (courtesy of LAEI).

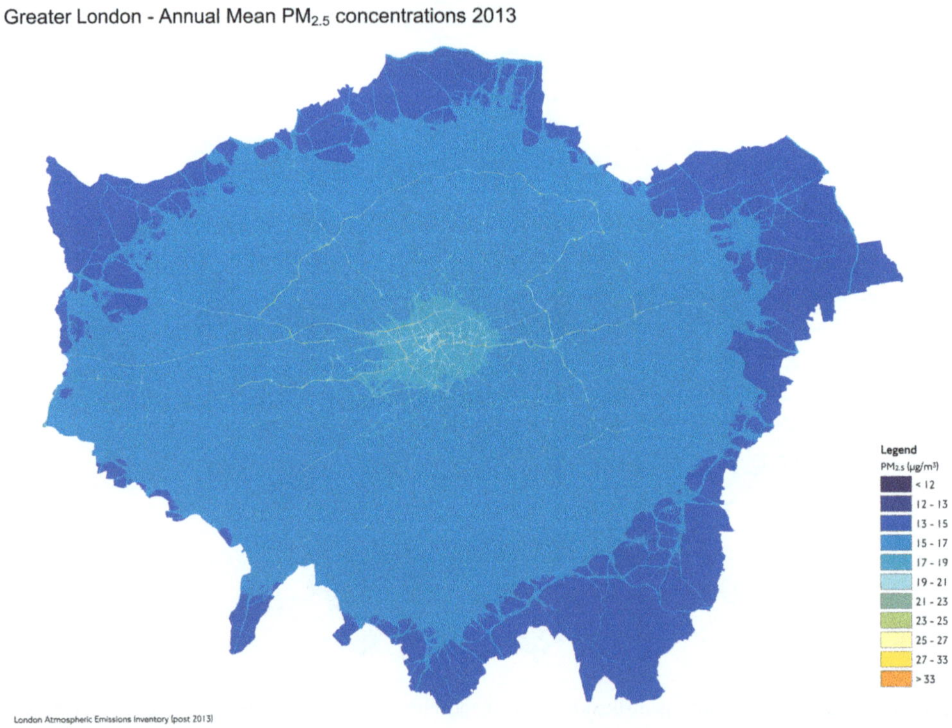

Figure 4.12. Annual mean concentration of $PM_{2.5}$, particulate matter less than 2.5 microns (µm) (courtesy of LAEI).

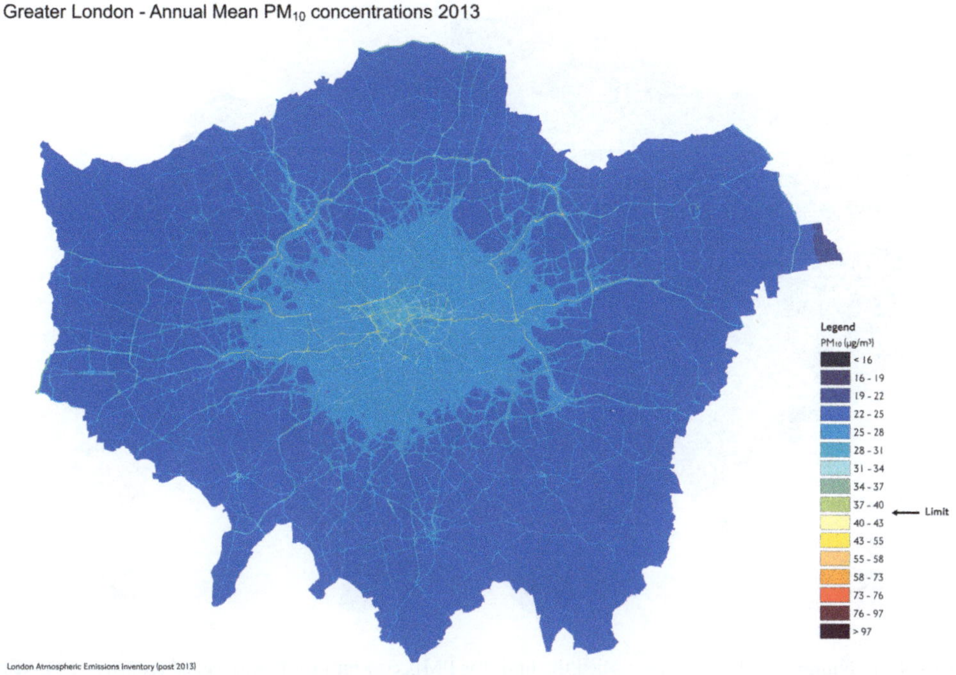

Figure 4.13. Annual mean concentration of PM_{10}, particulate matter less than 10 microns (μm) (courtesy of LAEI).

London Atmospheric Emissions Inventory (LAEI) and show the post-2013 distributions of various pollutants across the British capital (figures 4.10–4.13).

We can see from the maps that airborne pollutants are concentrated around the city centre, Heathrow airport (on the Eastern side of the city), and around major thoroughfares. The PM_{10} exceedance map (figure 4.14) shows that the consistently greatest concentrations of particulate matter are found along the major roads into and around London. This is not only due to the production of particulate matter by motor vehicles but also the trapping of these particles within street canyons.

Increasing vegetation is often touted as a way to reduce air pollution in urban areas. Gaseous pollutants such as NO_2, can be absorbed through stomata, however, the species of plant is important. Plant species with broad leaves are effective at taking up pollutants; however, these tend to be deciduous and will only have an effect during spring and summer. Evergreen trees, such as conifers, while active all year round are not as effective at taking up pollution. For particulate matter, vegetation is not very effective at all, with the deposition of particulates onto vegetation being small (~1% effect). Instead, in the absence of rainfall to remove particulate matter from the lower atmosphere, a varied landscape with increased surface roughness is required to disperse the particulates (and other pollutants). A varied landscape is aerodynamically rougher than a homogeneous one, leading to increased turbulence and vertical transport (figure 4.15). Parks and other green spaces will reduce the concentration of air pollutants not only through take-up and

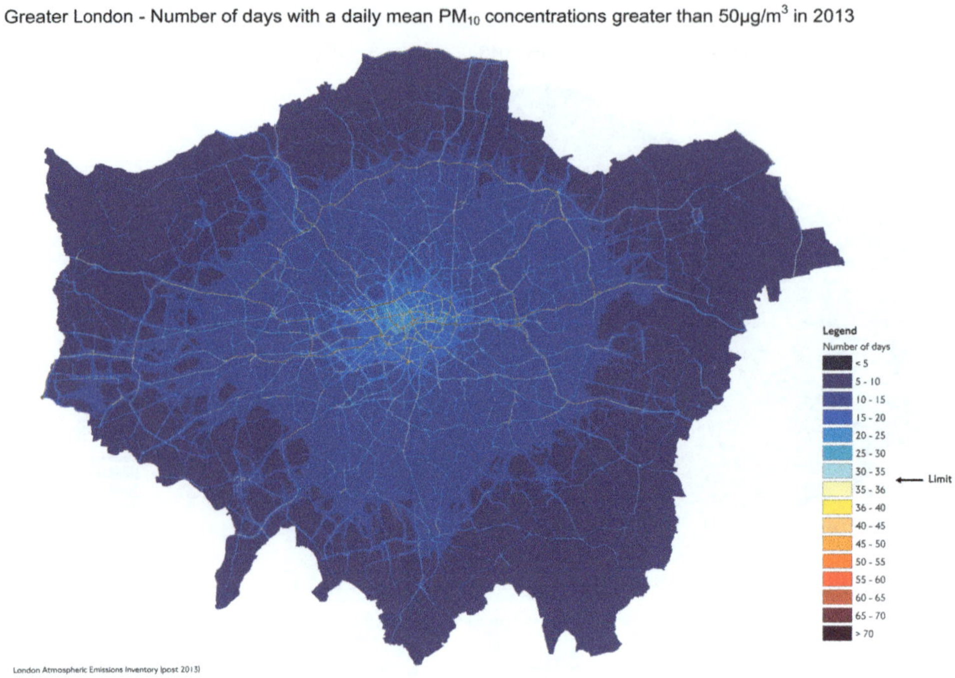

Figure 4.14. Number of days per year when the limit for PM_{10} concentration is exceeded (courtesy of LAEI).

Figure 4.15. Parkland and buildings of different heights provide greater surface roughness to disperse pollution (top) while less roughness concentrates pollution at ground level (bottom).

deposition, but also through increased surface roughness, increased evapotranspiration, and hence increased vertical transport (remember water vapour is lighter than air), and finally by providing a space into which pollutants can disperse via Brownian motion, thereby having a diluting effect on airborne pollution.

In the same way as green spaces can dilute air pollution, the urban geometry and trees lining roads or pedestrian walkways can alleviate air pollution. In figure 4.16, we are presented with two situations. In both cases, the air within the street canyon is isolated from the air above the UCL; however, in the top case the pedestrians are separate from traffic and the trees will act to filter the air within the street canyon. Conversely, in the bottom case, the presence of trees act to further isolate air in the canyon from the atmosphere above, and air pollution can become trapped and concentrated beneath the tree canopies, thereby accentuating the risk to humans.

The challenges of safeguarding humans against high temperatures and air pollution in urban areas are inextricably linked via the environmental conditions imposed by the UHI. This, however, means that these issues have common solutions related to urban planning and the aerodynamic roughness of the cityscape. Furthermore, the presence of parks not only dilute air pollution but also provide evapotranspiration and promote vertical transport of air.

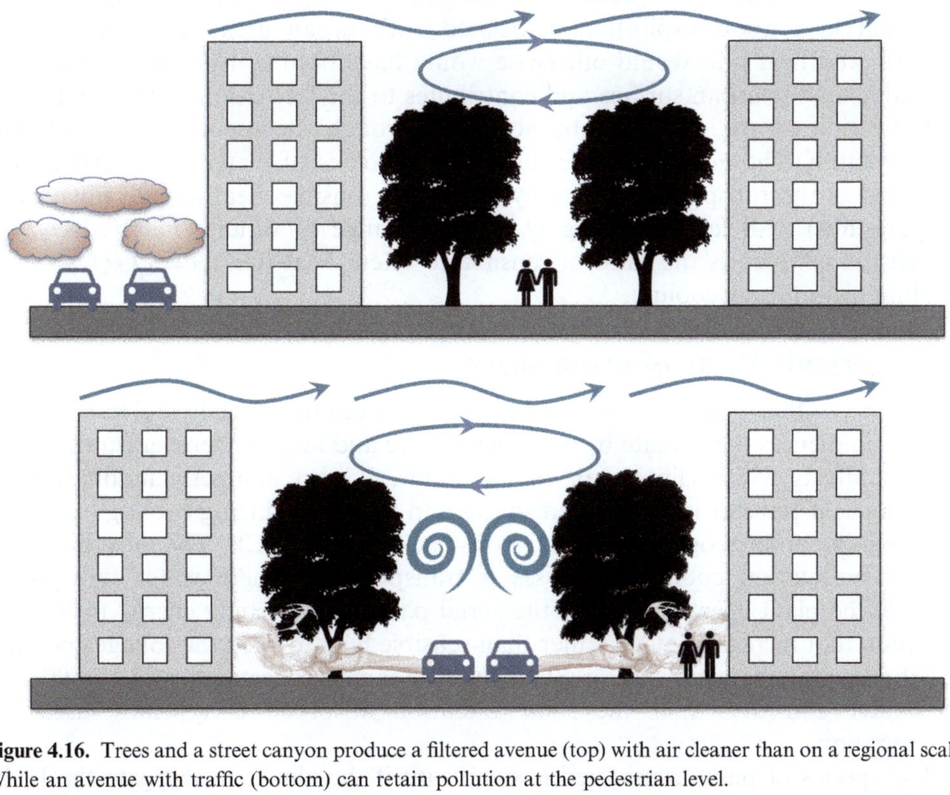

Figure 4.16. Trees and a street canyon produce a filtered avenue (top) with air cleaner than on a regional scale. While an avenue with traffic (bottom) can retain pollution at the pedestrian level.

4.6 Green and blue infrastructure

Green and blue infrastructure, or green and blue space refer to vegetated surfaces or bodies of water (static or dynamic) within an urban area. Green space can take many forms in an urban area, whether grassy parks or urban forests, tree-lined avenues or grass verges, private gardens, or the fringes of transport corridors. These green and blue spaces provide many varying ecosystem services to the urban area. These services include reduced surface runoff, flood relief, sustainable drainage, increased biodiversity as well as enhancing the general aesthetic and improving the health and wellbeing of the local populous. In this chapter however, we are primarily concerned with the ability of green and blue spaces to modify the local microclimate. A recent study of Glasgow (Scotland), suggested that an increase in green space around the city of ~20% could eliminate between 30%–50% of the city's UHI. The introduction of strategically planned and interconnected networks of green and blue spaces, offering ecological, social, economic, and climate resilience benefits are referred to in city planning as green or blue infrastructure.

The microclimate benefits of green and blue space arise from the evaporation and transpiration of water (collectively evapotranspiration), transferring energy from the surface to the atmosphere, thereby linking the urban energy balance with the hydrological cycle. It is estimated that the process of evapotranspiration consumes ~22% of the total solar radiation impinging the top of the atmosphere. The reduction in evapotranspiration associated with urban areas alters the energy balance, and heat that would otherwise would have been consumed as latent heat instead heats the urban surface and contributes to the formation of the UHI.

Mitigation of the UHI can be achieved through the use of green and blue infrastructure. The addition of vegetation and bodies of water to an urban area increases evapotranspiration, thereby replacing sensible heating Q_h (increasing temperatures) with latent heating Q_e (phase change of water). This reduces the *Bowen ratio*, which is the ratio of sensible to latent heat flux ($B = Q_h/Q_e < 1$), to produce evaporative cooling.

4.7 Thermal effects of green space

Green space affects the urban thermal energy balance both directly and indirectly. It directly influences the climate by reducing surface and local air temperatures, which in turn affects wider climate air temperatures. It also indirectly modifies it by reducing heat transfer into occupied spaces and thereby reducing cooling loads and any resulting anthropogenic heat emissions back into the UCL. The most discussed of such vegetation cooling processes is transpiration, where water transported through the plant is evaporated at the aerial parts by absorbing energy from solar radiation that increases latent rather than sensible heat to keep the foliage cool and thereby cooling the surrounding atmosphere. For most vegetated surfaces, 99% of the water and over half the solar radiation absorbed is typically used for transpiration.

The species of plant considered is significant if the aim is to mitigate the UHI. While C3 photosynthesis plants typically found in cool and wet climates transpire significant volumes of water, C4 photosynthesis plants that are adapted to warmer climates transpire much smaller volumes of water. Drought-tolerant plants typically found in arid climate use crassulacean acid metabolism photosynthesis and minimise water loss by keeping their stomata closed during the day, only opening at night. This effectively minimises daytime cooling potential due to their negligible transpiration rates. In the majority of plant types (C3 and C4), leaf stomata are typically closed in the absence of solar radiation (at night). Latent cooling from transpiration is therefore primarily a daytime energetic process. The rates of transpiration achieved during the day depend on the characteristics of the plant in question; crown area, leaf area index (LAI) defined as the single surface leaf area per unit of ground area, height of the leaves, stomatal resistance, the hydraulic resistance of the shoot and roots, and the local soil conditions described by dryness, compaction, and hydraulic conductivity, all have an impact on the transpiration rates achieved. The local climatic conditions also influence transpiration rates, as plants constrict or close stomata in order to control heat stress and water loss. A reduction in the cooling effectiveness of plants subsequent to protracted droughts or high temperature conditions is to be expected. This presents a problem when considering using green space to reduce the UHI and provide protection for urban areas against the higher temperatures expected as a result of climate change. The increase in temperatures and the shift in rainfall patterns from summer to winter will reduce the effectiveness of plants to provide evapotranspirative cooling. Furthermore, the use of drought tolerant plants from more arid climates, which is commonly cited as a suitable response will not solve this issue as these have a lower transpiration rate in the first place. At night and during periods of high temperature or when there is insufficient water, plants respire like other lifeforms rather than photosynthesise, emitting CO_2 and generating heat. Such effects are taken into account in global climate models as a feedback process, but this is given little consideration in urban climate discourse, which often opts for a blanket 'more green is better' approach.

As discussed in chapter 3, shading from vegetation keeps the atmosphere cooler through intercepting solar radiation and by reflecting energy due to its lower albedo or absorbing energy to drive photosynthetic processes and thereby limiting shortwave absorption by urban surfaces. This reduces urban surface temperatures and the eventual longwave radiative purging into the canopy-layer atmosphere. The reflection of radiation reduces shortwave absorption, with grass credited to reflect ~20%–25% and trees 15% of the incoming energy back to the atmosphere. A significant proportion of the absorbed shortwave energy is utilised by phyto-active chemicals in vegetation for biological photosynthesis, while the residual is held in storage. The effectiveness of vegetative shading on a surface is determined by the leaf size, crown area, and LAI of the vegetation canopy. As you might expect, trees, and to a lesser extent shrubs, are more efficient at shading than species of grasses. A tree canopy can therefore create a localised microclimate beneath it, so-called trunk-space cool spots (figure 4.17).

The canopies of vegetation alter the surface roughness, which in turn modifies wind flows and the convective heat exchange. The type of vegetation and the canopy density are particularly significant here, with grasses and low-lying shrubs creating a

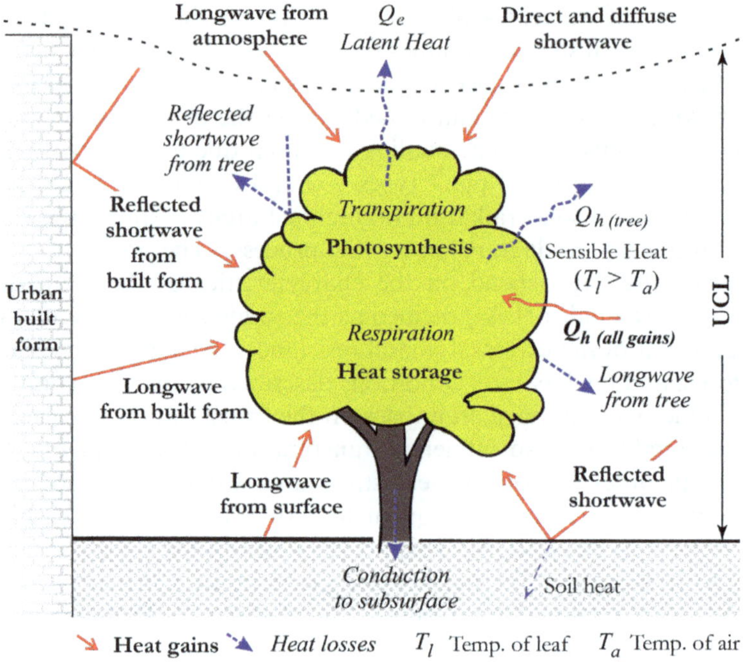

Figure 4.17. Energy exchanges between green space and the urban environment.

barrier of stagnant air near the ground and dense clusters of trees or forests impede wind flow to trap warmer insulated air beneath the canopy. However, dispersed groves of varied vegetation provide canopy heterogeneity, enhancing surface roughness, generating turbulence, and increasing convective heat loss. With lone isolated trees, the convective heat loss tends to be greater as they protrude through the UCL into the UBL, which increases exposure to drier air flowing from non-vegetated (urban) areas. This increases the water vapour gradient and increases evapotranspiration through leaf stomata.

In addition to these processes of evapotranspiration, solar shading, pollution filtering, and modification of wind flows, vegetation also indirectly cools the local climate. Particulate pollution filtering is achieved by dry deposition, a process where the pollutant molecules or particles impact upon and stick to vegetation surfaces such as canopy leaves. Removing particulate matter from the atmosphere reduces atmospheric scattering and absorption of shortwave solar radiation and longwave infrared radiation emitted by surfaces. This alters the radiation balance and the rates of atmospheric warming or cooling; typically, a clearer atmosphere will allow surfaces to warm faster, but also cool faster. As covered in chapter 2, vegetation canopies also reduce surface runoff by the interception of rainfall. The typical softer landscaping underneath vegetation also aids in the reduction of runoff rates and encourages greater absorption; this increases soil moisture, allowing for prolonged evaporative cooling after the rainfall event.

The cooling effect experienced depends upon the environmental conditions of the vegetated area. Background soil moisture content is particularly significant with precipitation or irrigation, providing greater soil water for transpiration, while high atmospheric humidity reduces the water vapour gradient and suppresses transpiration. Increased moisture availability also tends to determine the types and species of vegetation that result, with greater moisture availability resulting in denser growth, increasing surface roughness compared to drier climates. Wind flow is advantageous in high humidity conditions since it advects away accumulated saturated air, with higher wind velocities reducing the leaf boundary layer to enhance the water vapour gradient, increasing latent heat flux. These environmental variables of moisture content and wind flow, along with the ambient temperature, influence the types of vegetation native for a given area. This in turn determines the availability and effectiveness of the cooling processes discussed above.

In order to assess the thermal comfort and human health benefits of increasing green space, we need to have an understanding of the spatial extent of the impacts of any urban greening proposal. There have been many studies of the impacts of urban parks, but little consensus. A meta-analysis of several studies in the UK found that urban parks were on average 1°C cooler during the day, but that the extent to which this cooling permeated in the surrounding urban context varied considerably. A 1960s study of Kensington Gardens and Hyde Park in London found a 3 °C cooling effect which extended up to 200 m from the park boundaries. A more recent 2014 study of Kensington Gardens showed a mean summer temperature reduction of 1.1 °C with a maximum reduction of 4 °C observed on certain nights. The distribution of nocturnal cooling range was observed to be between 20–440 m, with 83% of the cooling potential still evident 63 m (the median of the range distribution) from the boundaries. The cooling influence of green space observed varies significantly both spatially and temporally. If vegetation is to be utilised for the cooling of urban areas, we need to understand the factors that determine the distribution of cooling (both horizontally and vertically) into the surrounding urban context.

Although numerous studies of urban parks have demonstrated the horizontal distribution of their cooling effects, there is little quantitative evidence to allow us to determine how isolated parks will affect the overall climate of a city. This issue is highlighted in studies of London. In contrast to most large cities, London is relatively green, with ~47% of its total area considered green, 33% vegetated green space, and 14% planted domestic gardens. However, a comparison of surface temperatures and the average atmospheric heat island for relatively warm summers (observations and simulation data output from the LUCID project, lead by University College London) reveals complex cooling distribution patterns (figure 4.18). Accounting for prevailing south-westerly winds, several areas of interest can be identified. Although cooling from Kensington Gardens and Hyde Park is apparent at the surface level, at the higher atmospheric level its influence is not apparent. Comparison of these images shows that combined green spaces of Richmond Park and Wimbledon Common, and to a lesser extent Hampstead Heath, are the only green spaces to have an influence in the higher atmosphere.

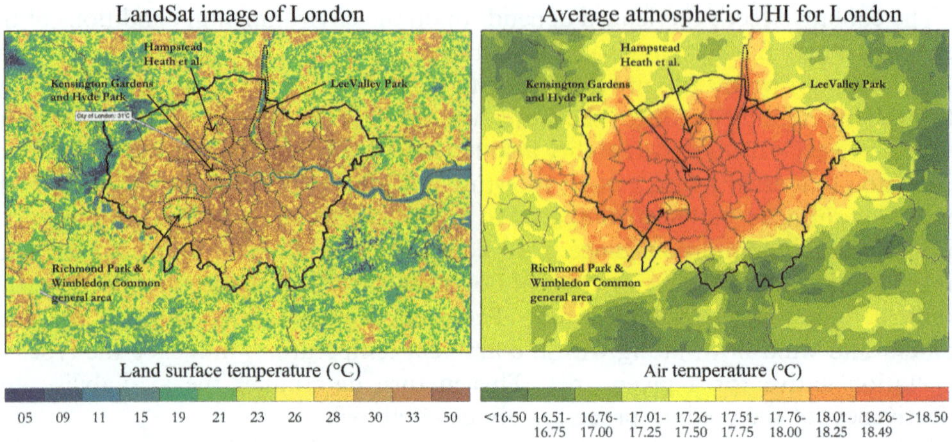

Figure 4.18. Surface temperatures from LandSat data (left) and UHI simulation data from the LUCID project (right)[2].

If we overlay the atmospheric (UBL) heat island data on top of a map of London showing green and blue spaces (figure 4.19), we can see that several large green spaces such as Hyde Park, Kensington Gardens, and Lee Valley Park appear to have no influence on temperatures in the higher atmosphere. There is also no visible influence from the river Thames or from other large blue spaces; this however, will be discussed later in the chapter.

From figure 4.19 we can see that the cooling influence of green space depends on more than just area. If pure size was the only variable in the cooling influence, then Lee Valley Regional Park would produce the greatest effect; however, its UBL cooling influence is not visible. It can be hypothesised that the magnitude and geometrical distribution of the green space has a significant bearing on the citywide (UBL) cooling expected. Table 4.1 shows the approximate dimensions of London's major green spaces. It is possible that the relatively linear geometry and limited fetch (for south-westerly winds) of the Lee Valley Regional Park (~1 km wide, compared with 5 × 7.5 km for Richmond Park and Wimbledon Common), impedes the development of the strong temperature and humidity gradients necessary to produce significant vertical transport and provide citywide cooling. Another possible contributing factor could be that a significant proportion (~22%) of this park is blue space, taken up by reservoirs. The significance of this is discussed later.

This hypothesis in relation to Hampstead Heath, which is relatively small, together with its surrounding context suggests that clustering of green spaces with suitable surface roughness is able to produce the vertical transport necessary to effect citywide cooling. There is currently a significant gap in the academic literature regarding monitored vertical distribution data that prevents testing of the relationship between geometric parameters and the vertical transport of cooling within the UBL. The majority of studies present data in relation to canopy-layer cooling, and almost entirely focus on the horizontal distribution of cooling. However, even in the

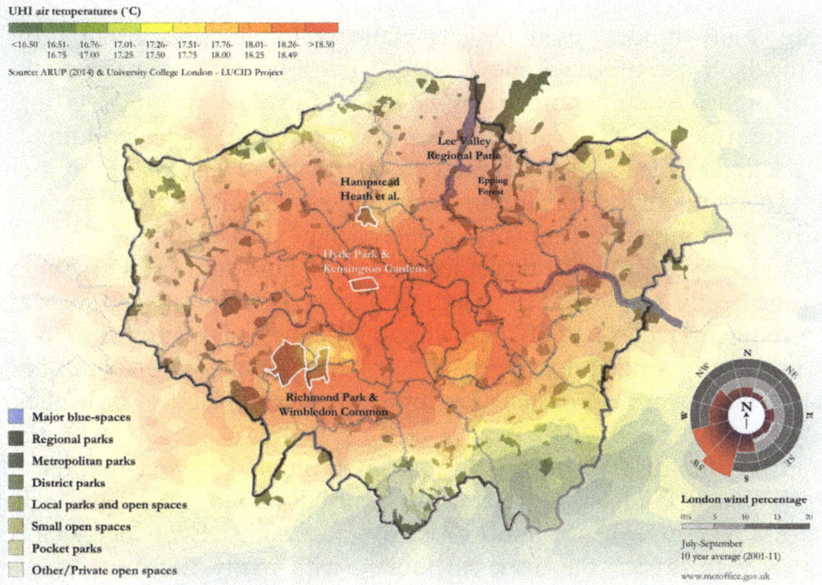

Figure 4.19. UBL UHI overlaid over a map of London's green and blue spaces[2].

Table 4.1. Major green spaces in London and their approximate dimensions.

Green space	~Area (km^2)	~East–west span (km)	~North–south span (km)
Hyde Park + Kensington Gardens	2.5	2.5	1.0
Hampstead Heath	3.2	1.7	1.8
Richmond Park	9.6	4.0	4.5
Epping Forest	24.8	2.7 (max)	8.8 (linear)
Lee Valley Regional Park	40.5	1.4 (max)	42.0 (linear)

absence of solid empirical evidence, it seems certain that the size, shape (determining fetch for different wind directions), and diversity of green spaces (determining surface roughness) influences the ability of a park to affect the UBL UHI and provide mesoscale (citywide) cooling.

[2] Details on the creation of figure 4.18 and figure 4.19 can be found in [2]

4.8 Green space implications for city planning

There are many models used by city-planners to describe urban growth with contrasting land-use morphologies and distributions. Of the many such models, the two theoretical extremes are the '*compaction*' and '*dispersal*' models (figure 4.20). Both feature widely in city planning discourse, but there is a general preference for the compaction model as it is believed to offer a more sustainable framework for growth. This preference for urban containment has its origins in a desire to safeguard the countryside from urban encroachment. The aim of the compaction growth model is to promote the efficient use of land and infrastructure. Higher density is generally considered to reduce resource consumption and lower economic and environmental costs (CO_2 emissions). The use of compaction policies are evident in modern England, where ~80% of the population occupies only ~20% of the land, with over 70% of new developments taking place on brownfield sites at relatively high densities.

In contrast, the dispersal growth model is promoted by free-market advocates who argue that if land-supply (greenbelt) restrictions are lifted, land will become more affordable and hence more socially accessible. Evolving socioeconomic patterns seem to favour the dispersal model, with many European cities moving from single urban centres to more poly-centric cities. However, the dispersal approach is treated with scepticism by policymakers, as allowing market forces to manage land-supply and demand may well lead to urban sprawl and low-density developments resulting in a high dependency on resource use. Furthermore, dispersal is associated with increased energy consumption, adversely affecting greenery and biodiversity, requiring extensive infrastructure networks.

Dispersed urban development is often criticised for increased land use compared to compaction or densification strategies, with urban growth likely to be on greenfield land and thus leading to the loss of peripheral green space. A US study showed that the rate of rural green space loss in the most dispersing urban areas was more than twice that in the most compact urban areas, with an association made between the frequency of extreme heat events and the loss of regional vegetation. Another study of Frankfurt (Germany) demonstrates the importance of safeguarding peripheral green space, showing a beneficial cooling effect of between 3 °C –3.5 °C. The Frankfurt study suggests that this cooling influence occurs due to the formation

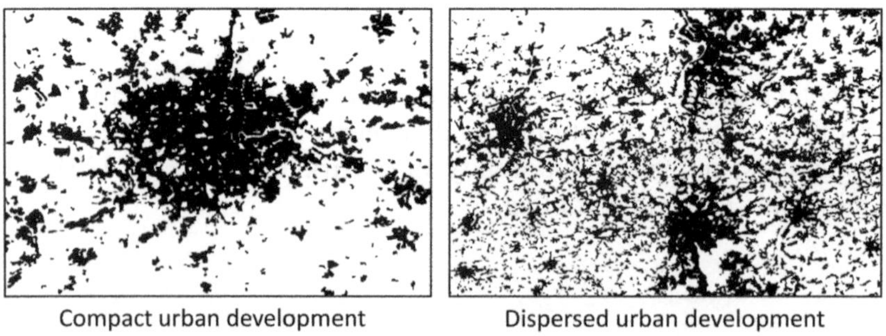

Figure 4.20. Illustration of the compaction and dispersal urban growth models.

of a mesoscale city–country breeze, which is also sometimes referred to as UHI flow. Under anticyclonic (e.g. warm, high pressure) conditions, this mesoscale system develops as thermals at the core of the city rise to the UBL, generating advection flow at the canopy-layer level from the cooler surroundings of the greenbelt (figure 4.21). Urban growth strategies that encroach peripheral green space can reduce this beneficial breeze by reducing the city–country temperature gradient, and hence the potential of the system, and thus prevent the supply of relatively cooler air from the greenbelt vegetation. In contrast, compaction strategies and urban densification actually enhance this phenomenon by enhancing the city–country temperature gradient. The amount of cooling expected or lost due to the loss of greenbelt vegetation depends upon the type and spread of the vegetation lost.

Until recently, reduced evapotranspiration in cities was considered the dominant factor for the formation of the daytime UHI. However, studies of US cities demonstrated that the daytime UHI is primarily dependent on the relative effectiveness with which urban and rural areas convect heat to the atmosphere, rather than evaporative cooling (heat storage is still the dominant factor for the night-time UHI). These studies found that the magnitude of the UHI was not directly correlated with precipitation or calculated potential evapotranspiration, but rather with convection efficiency. This study suggested that if urban areas are aerodynamically smoother than the surrounding rural areas (due to dense vegetation), heat dissipation is less efficient, resulting in warming; if the opposite is true, a cooling effect can be expected. The variation in convection efficiency between urban and rural locations is thought to be dependent on the background climate and its impact on vegetation cover in rural areas. In humid temperate climates in the US, convection was found to be less efficient at dissipating heat from urban areas than from rural areas, since rural areas tended to be aerodynamically rougher than the urban areas due to the presence of denser and coarser vegetation canopies. The study

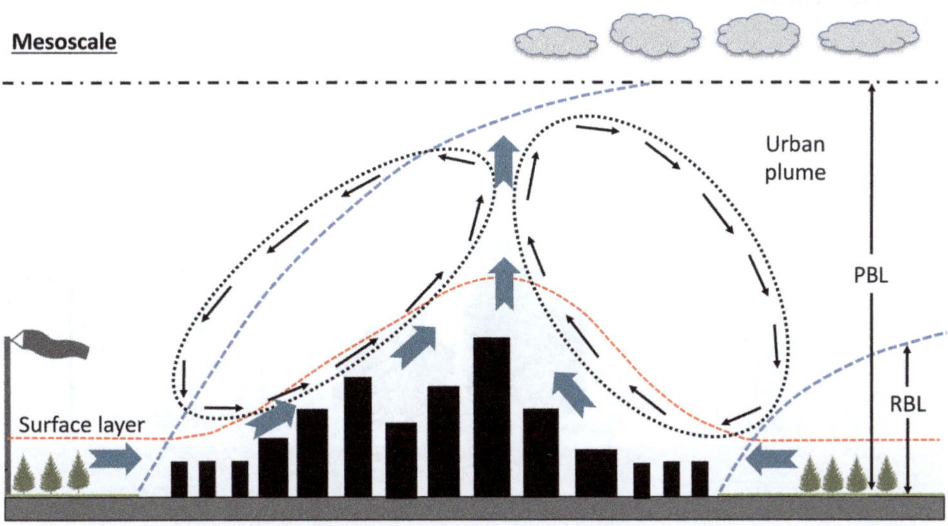

Figure 4.21. Schematic of the city–country breeze phenomenon.

found that US cities in such humid temperate climates showed a 58% reduction in convection efficiency compared to adjacent rural areas, resulting in a temperature increase of ~3 °C during the daytime. The opposite was found for US cities in drier climates, because the urban area tended to be aerodynamically rougher than the surrounding rural landscape, where the drier conditions inhibit the growth of dense vegetation. In such US cities a 1.5 °C decrease in daytime UHI intensity was observed. The study concluded, based upon climate model data and supporting field measurements, that the primary driver for the daytime UHI is the relative convection efficiency of the urban area compared to its surroundings rather than the reduction of evapotranspiration.

The formation of wind systems plays a significant role in the distribution of cooling from vegetated spaces. Macro-to-mesoscale prevailing wind flow and direction over the city determines the downwind spread, aided by a combination of simple advection along aligned street canyons and the turbulent mixing above the rooftops of cross canyons (perpendicular to wind flow). The formation of microscale wind systems has also been identified to play a significant role in the horizontal distribution of cooling from urban green spaces in a similar way to that from peripheral greenbelts. Under anticyclonic conditions, thermals rising from the surrounding urban area generate low-level advection currents that draw air from the park as 'park-breezes'. This park-breeze effect generates a centripetal convection current, which completes its cycle with the subsidence of warmer urban air from above into the park (see figure 4.22).

The presence of a park-breeze system may well explain why cooling rates observed within urban parks are seldom comparable to those of rural areas and are more closely associated with that of the surrounding urban area. The park-breeze phenomenon can also explain why many parks do not appear on heat island

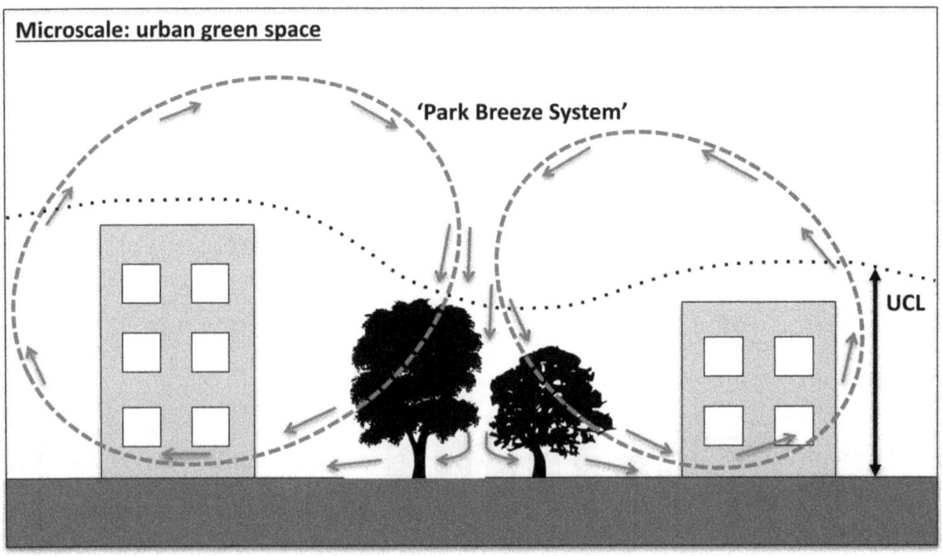

Figure 4.22. Schematic of the park-breeze phenomenon.

intensity maps such as figure 4.19. The presence of such a centripetal convection current hinders the vertical transport of cooling from the green space. The park-breeze system will only occur during periods of dynamic stability because higher wind velocities (>5 m s^{-1}) will create turbulent mixing and disrupt buoyancy effects. The low wind velocities typical of anticyclonic conditions, heatwaves, and high UHI intensity favours the formation of such a buoyancy-driven centripetal convection current. This implies that the UCL cooling influence of green space is enhanced by such microscale processes offering greater cooling distribution when it is most needed to relieve heat stress.

Studies have shown that cooling from green space penetrates further into the surrounding urban context for larger parks. The significance of scale can be attributed to the increased potential of the park-breeze system, either due to an increased temperature gradient or else due to increased fetch (length of area over which a given fluid flow has contact). The geometry of the park is important here, with square or round shapes providing greater cooling efficiency and distribution. This can be explained by a greater opportunity for increased temperature and humidity gradients and fetch between the park and the surrounding urban context. For a regular shape, the fetch will be similar irrespective of wind direction, making the formation of a park-breeze more likely. The range of the cooling distribution observed is also dependent on the types and diversity of vegetation. A recent modelling study combined tree age and planting density as a composite LAI to calculate the optimum cooling effect relative to park size. The results of this modelling supported the findings of previous studies in that networks of smaller 0.2–0.3 km^2 green spaces can provide effective cooling distribution. An earlier study that considered the scale and interval between bodies suggested that such network or cluster arrangements should be spaced <300 m apart in order to provide a combined benefit. There is however a minimum effective size of bodies to consider; green spaces smaller than 0.05 km^2 have been shown to offer negligible cooling. This reinforces the hypothesis that a certain fetch is required to generate a park-breeze system, and that larger parks are able to create larger park-breezes, allowing for greater cooling transport into the surrounding urban area, even for a constant temperature gradient. Currently it is uncertain if the same effect can be achieved through networks of smaller green-spaces, and what the required size and interval is in relation to surrounding urban surface roughness.

The findings of academic studies summarised above suggest that the addition of vegetation to an urban area with the principal aim of increasing evapotranspiration to mitigate the daytime UHI will have a minimal effect. At a mesoscale, the presence of vegetation seems to only aid the cooling of the city by enhancing its surface roughness. This provides another insight into the results presented in figures 4.18 and 4.19, where Richmond Park and Wimbledon Common presents a pronounced UHI reduction effect in contrast to Hyde Park and Kensington Gardens, not only because they are larger in area, but also because they are significantly rougher aerodynamically. It can be hypothesised that if urban greening is to be utilised for this purpose, tree planting with increased diversity of species will provide greater surface roughness than planar greening such as grassy parks. The type of urban greening to be used therefore requires consideration, not only of transpiration potential, but also of the resultant aerodynamic roughness produced by varied arrangements. Certain countries have developed planning processes that account for the relative environmental capital of different green space land cover. For example, the Green Area Ratio implemented in Berlin (Germany), and adapted in Malmo (Sweden), assign weighting factors based on their relative climate mitigation potential.

With regards to compact or dispersed development, we have a choice. Do we increase the density of our cities, with taller buildings enhancing surface roughness and increasing the city–country temperature gradient, and rely upon the city–country breeze to provide UHI cooling? Or do we have more dispersed developments with networks of green infrastructure to provide park-breeze UHI cooling? The former is more energy and infrastructure efficient, but the latter provides greater environmental capital. The best solution is currently unclear, and each urban area needs to be considered independently, weighing all the competing environmental and market forces.

4.9 Green building envelopes

One solution to allow urban densification, and compact urban growth and increase green space, is the use of green building envelopes, which can take the form of vegetated façades or roofs. There is an increasing number of advocates for vegetal architecture (incorporating vegetation into architecture) who wish for trees and shrubs to be incorporated onto buildings as either retrofits or replacements. Particular examples of such approaches include the Park Royal Hotel on Pickering Avenue, Singapore, the Kensington Roof Garden, London, and Bosco Verticale, Milan (Italy).

However, despite this pressure it is unlikely that such greening will be possible in already compacted urban centres. A study of Hong Kong acknowledged that with extreme urban centre densities and limitations on roof load bearing capacities, only planar greening solutions such as herbaceous sward green-roofs may be possible. This study argued that green-roofs are less effective than street-level vegetation for UCL cooling, particularly when typical building height exceeds 10 m. This is particularly true for Hong Kong, where the average building height is 60 m. Therefore, the UHI cooling influence of green-roofs was deemed negligible. While the cooling effects of vegetated roofs will not permeate to street level and directly reduce the UHI, there are the added benefits of reducing the albedo of the urban surface and the reduced heat penetration into the top floors of buildings, which will indirectly reduce the UHI. This has to be weighed against cheaper options, however, such as just painting the roof white (so-called cool-roofs). Some studies have demonstrated that in comparison to extensive green-roofs (primarily grasses or Sedum type plants), cool-roofs may result in lower annual (heating and cooling) energy costs. However, when considered against the intensive green-roofs that include larger vegetative cover and greater heterogeneity, the opposite may be expected. The overall energy saving has been shown to generally increase with increased vegetation LAI, and in the case of cooling-dominated buildings, it is highlighted as the critical parameter. In terms of economic comparison, cool-roofing strategies tend to offer better value than green-roofing, which typically has higher construction and maintenance costs. The use of such greening strategies needs to be assessed beyond their simple economic practicality, and consider other benefits such as thermal insulation, reduced rainwater runoff, carbon uptake, and the ecosystem benefits offered.

A recent review of studies of green-roofs, concluded that proximity is significant to the cooling experienced and suggested that there is limited vertical transport of air, likely due to limited aerodynamic roughness (particularly for extensive roofs). Since the cooling from green-roofs does not affect temperatures in street canyons and there is little vertical transport of air to affect the mesoscale UHI, we can conclude that green-roofs are an ineffective UHI reduction measure. That is not to say that green-roofs do not have a place in a modern dense city, since they have many other benefits beyond the evapotranspirative cooling of air.

An alternative to green-roofs are green-façades. At first glance these may seem to be more appropriate to urban cooling because they are located within a street canyon inside the UCL. However, studies of such vertical greening solutions found that their cooling influence did not extend beyond their immediate foliage zone. Therefore, the principle cooling attribute is to shade building façades and reduce surface temperatures. There are however, added benefits in that such vegetation can help remove air pollution and provide some cooling to ventilated buildings. For more intensive living-wall type green-façades, which include irrigation systems and can support larger plants, the range of cooling influence is more evident (although still restricted to <1 m). The technology behind living-walls is constantly evolving, and there are few *in situ* measurements published, thus making generalisations about this type of greening strategy risky at this time. For instance, the Rubens Hotel in London incorporates a 350 m^2 soil-based living-wall that also acts as a vertical sustainable drainage system. The evaporative cooling effect of this system is likely measurable and is no doubt beneficial in cooling the local air, but how far this cooling penetrates into the surrounding urban context is unknown. The impact of vertical greening on street canyon conditions, where surface temperatures and heat storage determine the nocturnal canopy-layer temperatures, has not yet been sufficiently studied. The cooling potential and delivery of other ecosystem services in dense urban settings can be considered as an emerging field of research, and we cannot at this time quantify their benefit, if any. Therefore, incorporation of green-façades and living-walls should be done on the basis of their better-known ecosystem benefits.

4.10 Thermal properties of blue space

Urban blue space refers to the general category of features that incorporate the presence of substantial bodies of static or dynamic surface water. In many cities, substantial blue spaces naturally exist as part of their geography (e.g. rivers). In London for example, the River Thames is a dominant feature, which together with other blue-space represents ~2.5% of the city's area. Like green space, blue space can also be created or managed specifically to provide additional ecosystem services and local amenities. These can include reservoirs, canals, SuDS and rainwater harvesting systems. City planners and architects have long considered waterbodies as a vital strategy in minimising urban heat stress. Studies focusing on the cooling benefit of blue spaces are fewer compared to those for greenspace, and typically focus on the daytime influence on urban temperatures.

A waterbody's (static) or watercourse's (dynamic) ability to modify surrounding air temperatures is dependent on its inherent properties (temperature, area, depth, waviness etc.) and its interaction with the surrounding climate. Most studies focus on the evaporative cooling ability of blue space, transforming sensible heat into latent heat through the production of water vapour. The thermal properties of high specific heat capacity and enthalpy of vaporisation (or latent heat) give water a high thermal inertia, allowing waterbodies to act as a thermal buffer playing an important role in moderating peak temperatures and temporal variations. For large waterbodies such as oceans, over 90% of the solar radiation received over a year may be used to evaporate water, and for smaller water bodies this energy conversion is still likely to be >50%. Although on average this translates to a lower Bowen ratio for smaller, shallower waterbodies, the evaporative flux is characterised by diurnal variations. For a large part of the morning, the absorbed solar radiation is mostly used to warm the water (increase heat storage). Towards the afternoon when the water surface temperature and the water-to-air vapour pressure deficit peak, a strong evaporative flux is generated. The energy stored within the waterbody is adequate to maintain this evaporative flux even throughout the night, albeit with diminishing intensity.

The surface reflectance (albedo) of a water surface determines its radiative exchanges. The albedo of blue space is low (~0.1) at low to medium angles of incident solar radiation typical of higher latitudes, and varies day-to-day with flow rate and waviness, the biochemical makeup, and the presence of suspended particles (turbidity). This means that the majority of incident shortwave solar radiation, which includes ultraviolet, visible, and near-infrared light is absorbed by the water, leading to warming of the waterbody. In most waterbodies, solar radiation cannot penetrate any further than ~10 m. As we discussed in chapter 1, water vapour is a greenhouse gas and is opaque to thermal infrared radiation, and liquid water is almost a perfect black body in the thermal infrared part of the spectrum. Therefore, longwave (or thermal) infrared (wavelengths 4–100 μm) radiative flux from the atmosphere or surroundings is almost completely absorbed at the surface with minimal reflection, while the outgoing longwave flux remains relatively constant throughout the day (for larger waterbodies), owing to the large thermal inertia and the limited diurnal variation in surface water temperatures. The fluidic properties of water allow absorbed radiative energy to be transferred within the waterbody by conduction, internal radiation, and convection currents, all of which contribute to the efficient heat transport and mixing within the waterbody. These processes allow heat to be efficiently diffused throughout a large surface volume, maintaining surface water temperatures within a small diurnal range.

Higher surface water temperatures enhance the water-to-atmosphere temperature gradient. The temperature of the air above the water surface determines the sensible heat flux, with warmer air suppressing radiative heat loss, and cooler air enhancing the water-to-atmosphere temperature gradient, and hence increasing radiative flux. The water-to-atmosphere moisture gradient, also called the vapour pressure deficit, determines the potential for moisture to transfer to the atmosphere. Relatively drier air above the waterbody enhances evaporative flux as the vapour pressure deficit is increased, while more humid air above water surface suppresses evaporative flux. Increasing the wind velocity above a waterbody significantly alters both the sensible and evaporative heat flux by advecting away heat and moisture, enhancing both the temperature and moisture gradients. As with green spaces, high wind velocities can hinder cooling distribution by enhancing atmospheric mixing.

The thermal properties of dynamic (moving) watercourses, such as rivers, streams, and canals, are influenced by fluid flow variables (e.g. volume flow rate) and local climate parameters. The flow of water allows watercourses to carry by advection absorbed radiation downstream and release this energy in a different location, potentially outside the urban area. Measurements of watercourses have shown a marked increase in water temperatures downstream of urban areas, with temperatures shown to increase with increasing impervious surface area and surface run off in the urban area. A study of urban streams in Long Island (USA), found water temperatures to be 5 °C–8 °C warmer in summer and 1.5 °C–3 °C cooler in winter than rural streams. This study also noted that the diurnal temperature variation was greater in urban streams. During the summer, stormwater runoff from heated impervious surfaces leads to increases in the temperature of urban streams of 10 °C–15 °C compared to rural streams. Although such stormwater runoff has a beneficial cooling influence on the upstream urban surface, it can lead to thermal pollution and resulting biochemical issues, such as algae blooms further downstream.

The energy balance of a river is dominated by the net shortwave radiation balance, followed by contributions from the net longwave radiation balance and evaporative flux. For example, a study of River Exe in Devon (UK) showed the net radiation balance to account for 56% of the heat gain and 49% of heat loss. Since most watercourses are of limited depth for most of their course, thermal exchanges at the riverbed–water interface also require attention. Another study, this time of the River Blithe in Staffordshire (UK) found that 82% of energy exchange occurred at the air–water interface, while only ~15% occurred at the riverbed–water interface. For smaller streams, factors such as groundwater exchange and discharge from external sources have a greater influence on the total energy exchange, while for larger dynamic bodies, exposure to solar radiation and wind flow lead to the dominant heat exchange occurring at the air–water interface. This means that their water temperatures typically depend on the climatic conditions above and their diurnal and seasonal cycles.

Since the limited flow of static waterbodies restricts dynamic instability, they demonstrate higher sensitivity to energy exchange modifications at the air–water interface, meaning that their thermal properties are highly dependent on the climatic conditions above. For deep static bodies such as large lakes and reservoirs, thermal inputs can lead to temperature and density changes that produce strong thermal stratification of the water column. In such stratified waterbodies, the *epilimnion* describes the warmer upper layer; followed by the *metalimnion* or *thermocline*, where the temperature rapidly decreases with depth; and the cooler, denser, and more stable *hypolimnion* is at the bottom of the waterbody. This stratification of deeper waterbodies means that the thermally active zone for energy exchange with the atmosphere above is restricted to the epilimnion and the higher regions of the thermocline. There are several mixing mechanisms that can affect the thermal moderation of this active zone. At the surface of a waterbody, evaporative mixing can be observed, and since evaporation is an endothermic process, the latent flux generates instability that brings warmer water to the surface to maintain a relatively constant surface temperature. Of far greater significance is the mechanical energy transferred by wind shear stresses at the water surface, producing waves and turbulence. The strength of this mixing is dependent on the characteristics of the wind flow at the waterbody's surface, as well as the available fetch and the presence of any littoral (riverbank/shoreline) obstructions. Strong wind-driven turbulence in the waterbody may transfer turbulent kinetic energy to the lower layers, leading to a breakdown of the stratification. In addition to these forms of surface mixing, a diurnal mixing current can be generated at the littoral zone (near the banks) of a waterbody. Since the shallower water on the littoral slope heats faster than the open water during the day, a horizontal current is generated from the littoral zone towards open water, while cooler water from the open water depth is drawn up the slope. This current is reversed at night because the littoral zone cools faster than open waters. This diurnal littoral zone current, however, is generally not strong enough to disrupt the stratification of the entire waterbody.

Whole waterbody mixing is evident in holomictic waterbodies common to temperate climates, which have a uniform temperature and density from top to bottom during certain seasons. The seasonal changes in temperate climates produce the conditions necessary for buoyancy-driven overturning of the stratification structure. This overturning occurs at a threshold temperature of ~4 °C, when water reaches its maximum density. During the spring, when surface water that is cooler than this threshold temperature starts to warm, its density increases and falls through the waterbody creating a convective instability. This continues until the epilimnion reaches ~4 °C, after this point, further warming increases stability and restricts vertical mixing, leading to stratification. The reverse is true in autumn: as the surface water cools, its density increases, again creating convective instability and the mixing of the whole waterbody. Waterbodies that demonstrate this type of biannual mixing are referred to as dimictic bodies, and waterbodies in cold winter climates have been observed to demonstrate rapid overturning. Apart from this seasonal overturning, all other natural mixing processes act to maintain the stratification structure for most of the year. During these periods, only the epilimnion and the upper thermocline act in energy exchange with the climate above, since the stratified state of the waterbody prevents the use of its entire volume and thermal capacity.

For shallower static bodies, their limited volume places a limit on the available thermal capacity and inertia. Therefore, the surface temperature and resulting latent flux peaks earlier (daily and seasonally) than for deeper bodies. The reduced volume of water increases the significance of the processes discussed above, such as littoral zone and wind-driven mixing. Furthermore, the conduction of heat across the water-bank boundary into the surrounding ground, and the absorption of heat by aquatic flora and fauna, are likely to be more pronounced. The increased effect of these energetic processes means that for shallower waterbodies the net radiation balance converted into evaporative flux is lower than for larger ones, but still a substantial value. For instance, a study of a shallow lake in the Hudson Bay lowlands (Canada) showed that on average 55% of the daily net radiation balance was still utilised for evaporative flux. The reduced depth of such waterbodies allows shortwave radiation to penetrate the full depth, allowing solar radiation to be absorbed (or reflected back) and heat to be conducted into the subsurface. This can lead to water-to-bed interface energy exchanges and possible warming of the water from below. Shallow waterbodies are generally considered not to exhibit stratification of the water column as a result of wind-driven and evaporative mixing; instead, shallow waterbodies are often considered to be a well-mixed epilimnion, able to use their entire volume and thermal capacity for energy exchange. However, there is evidence that even shallow waterbodies can demonstrate stratification during prolonged periods of warm calm weather typical of heatwaves and high UHI intensity, which reduces their effective thermal capacity and limits air–water energy exchange when it would be most beneficial.

Smaller and shallower ponds have become more common place for SuDS, which is now common practice within urban development plans. The urban setting has a profound effect on the thermal properties of such ponds. The inflow of warm summertime surface runoff as well wind sheltering from surrounding buildings inhibiting wind-driven surface mixing means that urban ponds behave different thermally from their rural counterparts. These urban influences may result in the development of thermal stratification in such ponds. For example, a study of ten shallow urban ponds in Ontario (Canada) showed that the density variation produced by daytime heating was not always fully dissipated at night depending on the surrounding environmental and subsurface conditions. The resultant stability and stratification extended through the mid-summer months, with the ponds demonstrating vertical temperature differences of ~4 °C between the top and bottom layers. This strong and persistent stratification in a shallow pond was attributed to the turbidity of the water, which contained high levels of suspended sediment and organic particles, and the reduced wind stress observed at the water surface.

In terms of geometric parameters, the ratio between the water's surface area and perimeter (shape factor) and the maximum depth are significant. Examination of the influence of the shape factor reveals that shallow ponds with a relatively large area but simple geometry (round or square) experience pronounced stratification as opposed to larger lakes where larger surface areas with greater fetch are typically associated with greater turbulent mixing. The maximum depth unsurprisingly demonstrates the greatest correlation with stratification, with only the shallowest waterbodies <1 m deep identified as being relatively isothermal (a well-mixed epilimnion). This means that only very shallow waterbodies are able to utilise their entire volume's thermal capacity for air–water energy exchange.

4.11 Thermal effects of blue space

As with green spaces, the cooling effectiveness of blue space in terms of magnitude and distribution is dependent on, amongst other things, the size and spread of bodies, and the distance from them. A recent mesoscale modelling study of a hypothetical waterbody situated within an idealised city showed that relatively large waterbodies demonstrate greater cooling, particularly close to their boundaries and in downwind areas. The length of the downwind distribution was dependent on wind velocity, with relatively cooler air from a large urban waterbody able to create plumes several km long. This modelling study also confirmed previous findings that several smaller regularly shaped waterbodies distributed evenly throughout the urban area have a smaller cooling effect (particularly during the day), albeit across a greater area.

A study of Beijing (China) observed that waterbody geometry is significant for cooling distribution, with square or round geometries providing greater cooling efficiency. As discussed previously in relation to green space, this can be attributed to the increased temperature and moisture gradients that are likely to result between such wider waterbodies and their surrounding landscape. Additionally, regular geometries present a more consistent fetch for wind irrespective of wind direction, presenting greater opportunity for atmospheric advection. Similarly, the Beijing study highlighted the importance of a watercourses width, with urban river width being a principal factor affecting the temperature and humidity of the riparian (area adjacent to rivers) zone. The study found that when a river is >40 m wide, there is a significant and stable effect of decreasing temperatures and increasing humidity, indicating the transport of air away from the water's surface. This may explain why the river Lea in the Lee Valley Regional Park, which is <30 m wide through much of the park, does not appear to have a cooling effect on the atmospheric UHI for London (shown earlier in figure 4.19).

An urban setting modifies the climatic variables that affect a waterbody. For instance, higher water surface temperatures are observed for waterbodies surrounded by buildings that store and radiate heat. This creates a greater temperature gradient between the centre of the waterbody and its surrounding context, enhancing cooling distribution. The surrounding urban morphology further influences this cooling distribution by shading the waterbody and obstructing wind flow. Shading alters the net radiation balance, reducing the temperature gradient and evaporative flux; meanwhile, wind sheltering reduces the opportunity for atmospheric advection and surface layer mixing from waves. Littoral sheltering, whether from buildings or vegetation, reduces the wind stress induced turbulent mixing (as does aquatic vegetation), resulting in a reduced surface layer depth (and reduced volume or energy exchange) and potential strengthening of temperature stratification. The degree of mixing generated depends on both the amount (cover) of sheltering present and the available fetch. The longer the fetch, the greater the opportunity for turbulent mixing. For larger lakes with increased fetch, littoral sheltering damps surface waves, preventing turbulent kinetic energy from penetrating into the water column. The effectiveness of this damping is dependent on obstruction separation, with higher obstruction density achieving greater damping. Meanwhile, as obstruction separation is increased, turbulence also increases until it is similar to open water.

Farms in Australia often have such obstructions upwind of small reservoirs to reduce the evaporative water loss. A study of the effect of these littoral windbreaks (both artificial and natural) on the evaporation from these small waterbodies showed that reduced wind speeds and increased air turbulence immediately downwind of a shelter lead to the accumulation of moisture in the air immediately above the water surface, reducing the humidity gradient and evaporative flux. The area immediately downwind of the littoral obstruction showed reduced waviness of the water surface, and hence the water's surface area in contact with the atmosphere is also reduced; this is termed the *quiet zone*. Wind speeds downwind of the obstruction decrease to a minimum at ~3× the height of the obstruction (see figure 4.23). While the point of minimum wind speed does not coincide with the edge of the quiet zone, the two are related, as shown below.

$$x_{\text{quiet}} = 1.5 + 0.92x_{\text{min}}.$$

Wind speeds (u) farther downwind of the obstruction remain reduced, although turbulent intensity increases as wind flow recovers to that upwind of the obstruction. This turbulent *wake zone* exhibits increased moisture and heat transfer. The study found that greater wind protection reduces the wind speed experienced by the waterbody, and therefore reduces evaporative cooling; however, a corresponding increase in surface water temperature was not observed. It can be hypothesised that this is due to increased radiative output, or increased conductive heat loss between the water surface and the moisture saturated air that forms above the water surface.

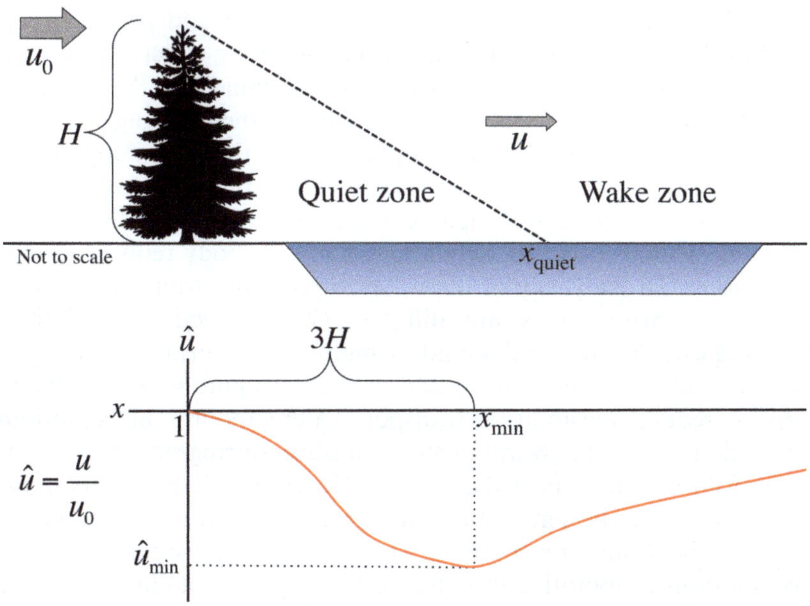

Figure 4.23. Sheltering effects of littoral obstructions on a waterbody and resultant wind velocities.

Finally, the study found that evaporation from the water's surface varied throughout the day with a distinct diurnal cycle. At night when the Richardson number (the ratio between buoyancy and flow shear, or equivalently, the ratio between natural and forced convection) was suggestive of atmospheric instability (see figure 4.5), the sheltering effect of a windbreak was reduced.

The sheltering effect of natural (e.g. trees) and artificial (e.g. buildings) littoral windbreaks is another possible reason why Lee Valley Park, which contains several lakes and reservoirs, does not appear to affect the UHI shown in figure 4.19. Due to the linear geometry of the park and surrounding trees, the water surface may well be completely sheltered (i.e. entirely within the quiet zone), reducing the cooling influence of the blue spaces. Even within the wake zone, where evaporation is increased due to air turbulence, the same turbulence will inhibit the vertical transport of cooling. This implies that thought has to be given as to the placement of blue space in relation to the prevailing wind direction and any surrounding littoral windbreaks, so as not to create localised stagnant areas that exhibit higher night-time temperatures (due to the large thermal inertia of water) and higher humidity (from moisture saturated air), which can be damaging to human health.

The blue space cooling distribution demonstrates distinct diurnal and seasonal variations. For example, a study of an urban river in Sheffield (UK) observed that the cooling effect tends to be greatest in the morning, with warm days in spring demonstrating ~2 °C lower temperatures over the river and ~1.5 °C in the riparian zone. No night-time cooling was observed. Towards late June (mid-summer) daytime cooling had notably diminished, with similar air temperatures observed over the water and elsewhere. In agreement with such observations, the mesoscale model of a hypothetical city mentioned above, showed blue space cooling as principally a daytime phenomenon; meanwhile, at night and particularly towards the mid/end of the summer, a distinct nocturnal warming effect can be expected.

These diurnal and seasonal variations can be explained by the variation in the evaporative flux. The instability of the atmosphere in the morning will increase the evaporative flux into the immediate atmosphere above the waterbody, increasing its water vapour content. In the early afternoon, atmospheric convective instability peaks along with the water surface temperature and evaporative flux, although the vapour content of the air immediately above the waterbody reduces as the surface warms, increasing the buoyancy of the water vapour and transporting it to higher altitudes where concentrations are diluted. The convective instability of the atmosphere reduces during the day and at night as the surface atmosphere cools, gaining stability and resistance to vertical transport (figure 4.5). The stabilising of the atmosphere reduces the ability to transport water vapour to higher altitudes, and thus evaporation from the warm water's surface during the evening leads to increasing moisture content near the surface. Hence, the stable layers immediately above the water surface become saturated with evaporated water vapour, reducing the moisture gradient and preventing any further evaporative flux.

The reduction in evaporative flux during the night results in a warming of the water's surface (increasing radiative output). The differential cooling rates of the waterbody and its surrounding urban context (which cools faster), reduces the waterbody-to-context temperature gradient. This reduces the likelihood of nocturnal horizontal advection currents, which could vacate the saturated stable air mass above the waterbody. This warming effect is particularly pronounced when waterbodies reach higher temperatures towards the end of the summer from accumulated thermal energy. Simulations of the diurnal cycle of a waterbody indicate a cycle of warming and cooling as follows:

- Morning (~10 am)—cooling at low altitudes (<500 m), warming at higher altitudes.
- Afternoon (~3 pm)—modest cooling at all altitudes, greatest vertical transport.
- Night-time (~4 am)—warming but at low altitudes only, minimal vertical transport.

It is possible to hypothesise that the differing thermal properties of waterbodies and the surrounding urban context generate distinct microscale breeze systems, similar to those observed above green spaces. Such a waterbody-breeze system would differ from a park-breeze system due to the high thermal inertia of water and its diurnal cycle. The waterbody-breeze system could thus be expected to reverse during the night as warm saturated air rises from the warmer waterbody causing cooler air from the urban surroundings to advect horizontally towards the waterbody (figure 4.24). The completion of this waterbody-breeze centripetal cycle would be the night-time subsidence of warmer and humid air back to the surrounding urban area. This presents the possibility for the horizontal transport of an undesirable warming effect into the surrounding areas. The canopy-layer trapping of heat and moisture presents significant threat to not only thermal comfort, but also to human health and wellbeing.

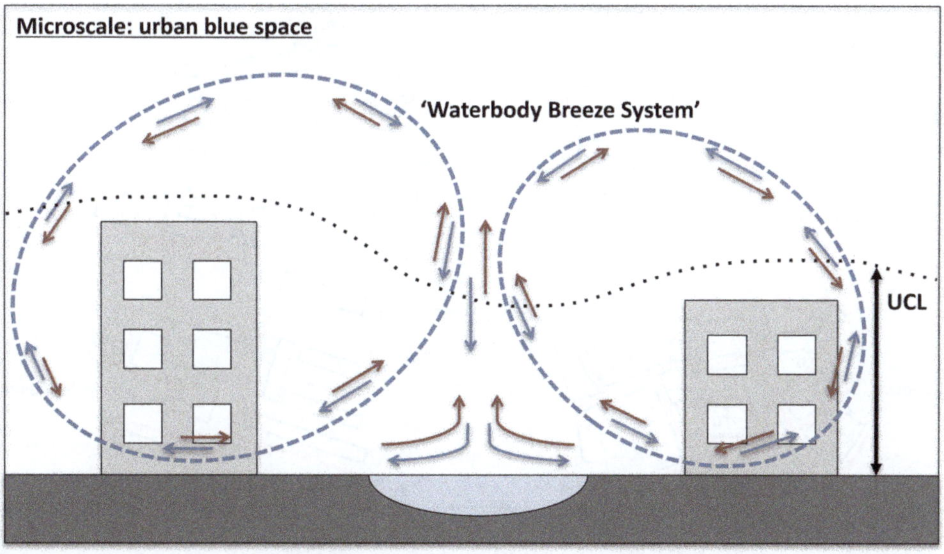

Figure 4.24. Schematic of the waterbody-breeze phenomenon.

The suggested cooling benefit of evaporation from waterbodies is somewhat misleading, even during the day. A significant drawback of evaporative cooling is that it increases the water vapour content of the air (humidity). Increased humidity reduces the effectiveness of human thermoregulation by reducing the readiness with which sweat can evaporate, but also, since water vapour is a greenhouse gas, by altering the emissivity of the surrounding air leading to greater absorption and re-radiation, trapping heat at street level. Studies have observed that up ~60% of the cooling benefit (to humans) from reduced air temperatures (sensible cooling) can be negated by the increase in humidity. Consideration of the diurnal thermal cycle of blue space suggests that it warms the urban environment when it is least desirable; at night, and particularly under anticyclonic conditions typical of high UHI intensity and heatwaves. Hence, we can conclude that blue space offers limited potential for UHI mitigation if considered in isolation.

4.12 Urban planning for the UHI

The design of urban areas, particularly those in warmer locations, has evolved over time to reduce the UHI, citywide temperatures, and heat stress. This largely takes the form of buildings providing mutual shading or the alignment of street canyons to allow cooling breezes to penetrate deep into the city. With freely available satellite images, it is now relatively easy to compare urban layouts for different cities around the world. Figure 4.25 shows real streets (red lines) and building (solid black lines) layouts for sections of four cities at the same scale using data obtained from ArcGIS24. The buildings shown are primarily domestic to allow a fair comparison.

We can see from that for cool maritime climates such as London and Bath, buildings have a compact form, typically with small openings to minimise heat loss.

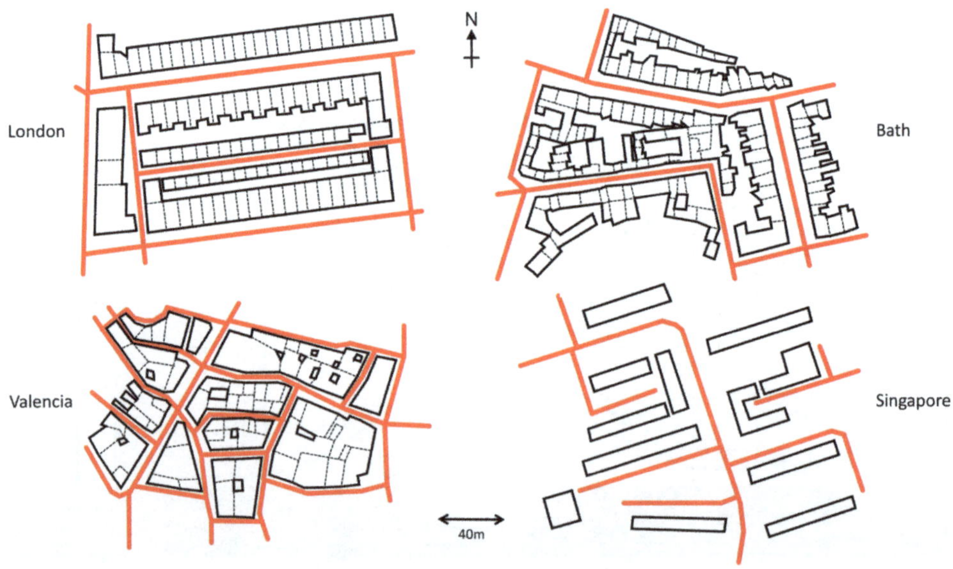

Figure 4.25. Example urban layouts for different cities, showing buildings (black) and roads (red).

For warm dry climates such as that of Valencia, the urban form is more compact, and buildings are closer together to provide mutual shading at the lower latitude, with central courtyards used to create a local microclimate (a well of cool air) used to cool the inside of the building. As such, the urban street canyons have a greater height to width ratio than in cooler climates, reducing the UHI by limiting solar gain during the day. Buildings are typically light-coloured to reflect more shortwave solar radiation. We can also see from figure 4.25 that the urban form of Singapore is sparse, with free spaces between buildings to promote airflow through the urban area. Another distinguishing feature between the urban form of the four cities is the amount and placement of vegetation. For London and Bath, there is some grass and a few trees located in the back gardens of the homes. For Valencia, there is no grass visible in the satellite image, but there is a greater number of trees, which line the roads and shade the façades in the street canyons. By comparison, satellite imagery of the area of Singapore shows more green space than man-made surfaces. There are a large number of trees with grass beneath, located between the buildings, shading the façades. The large amount of vegetation in this urban area and the sparse urban form effectively minimise the UHI. These are just some of the architectural traits that have evolved over time in different locations to control temperatures in buildings and urban areas.

As you move to hotter drier climates, such as the middle East, the urban form becomes more compact, which you may expect to prohibit the ingress of wind and the transportation of heat away from the urban area. However, in such areas the urban layout has evolved so that street canyons are aligned with cooling night-time breezes to purge heat. This is possible because in continental locations with reduced topography, wind direction is generally very predictable, with winds only coming from a small range of angles. This can be seen in figure 4.26, which shows radar plots of wind direction and speed, or *wind roses*, for two cities in Saudi Arabia.

Aligning street canyons roughly east–west will allow wind to flow through these street canyons, allowing the urban fabric to be cooled rapidly at night. This is in contrast to the UK, which, as shown in figure 4.27, has a much less predictable wind direction. Less predictable wind directions makes it much harder for to design urban areas to allow consistent wind ingress and to shed heat at night.

Another way in which the design of urban areas can be adapted to reduce the UHI is by maximising the sky view factor of buildings and façades. The sky view factor is a dimensionless value that ranges from 0 to 1, and is the fraction of sky visible from the ground. The denser or taller buildings are, the lower the sky view factor (figure 4.28).

If we remember back to earlier in this chapter, we recall that although shortwave solar radiation can be reflected by surfaces, conserving momentum, until it is absorbed or leaves the atmosphere, longwave thermal infrared radiation given off by surfaces is radiated in all directions equally. This means that for a warm surface, the sky view factor can be viewed as the probability that the radiation will leave the street canyon and pass out to space. The lower the sky view factor, the more likely the thermal radiation will be absorbed by another urban surface, contributing to the UHI. The sky view factor of a given street canyon or building façade can easily be

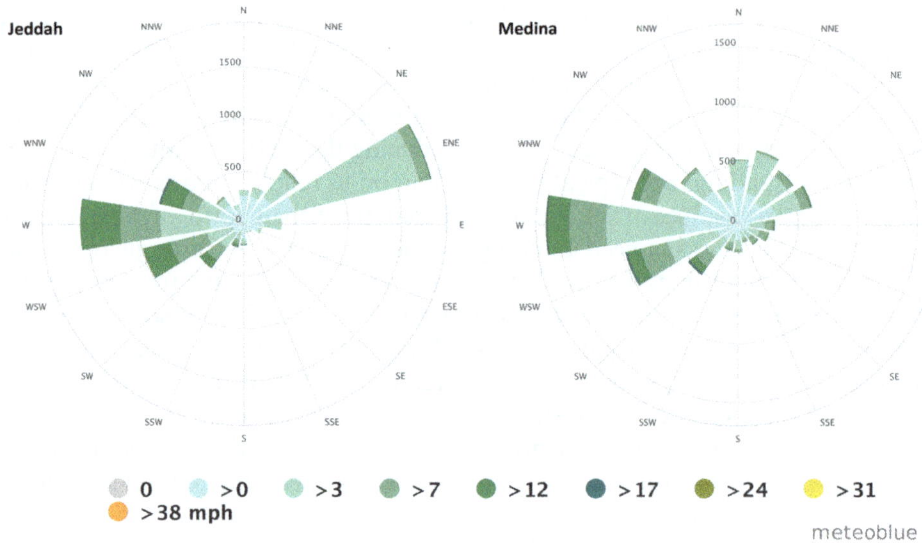

Figure 4.26. Wind roses for two cities in Saudi Arabia (Jeddah and Medina).

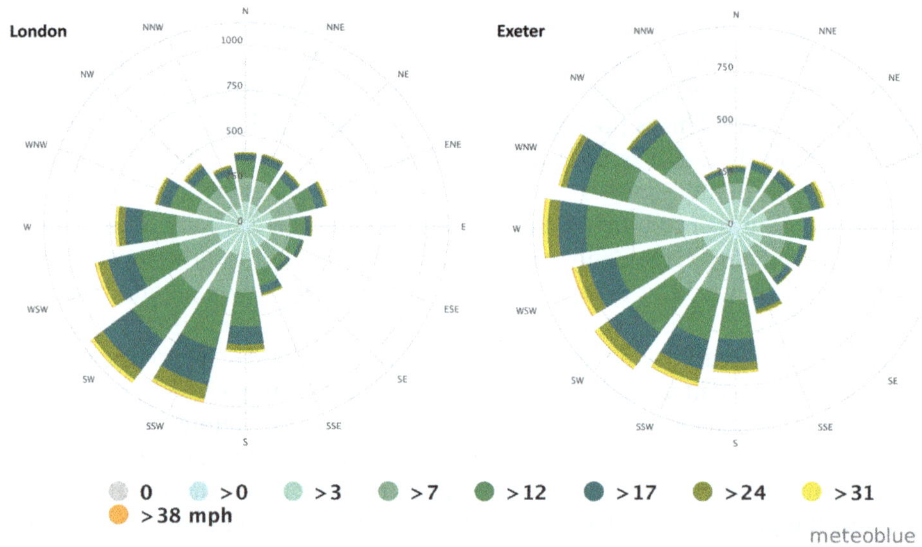

Figure 4.27. Wind roses for two cities in the UK (London and Exeter).

assessed with the use of a camera equipped with a fisheye lens, which can see all the directions in which heat could be radiated. The sky view factor is then simply the percentage of sky visible in the image, as is shown in figure 4.29.

The technique of maximising the sky view factor as part of urban planning has, for example, been implemented in Ho Chi Minh City (Vietnam). In Ho Chi Minh green spaces are surrounded by small buildings, with medium-sized buildings

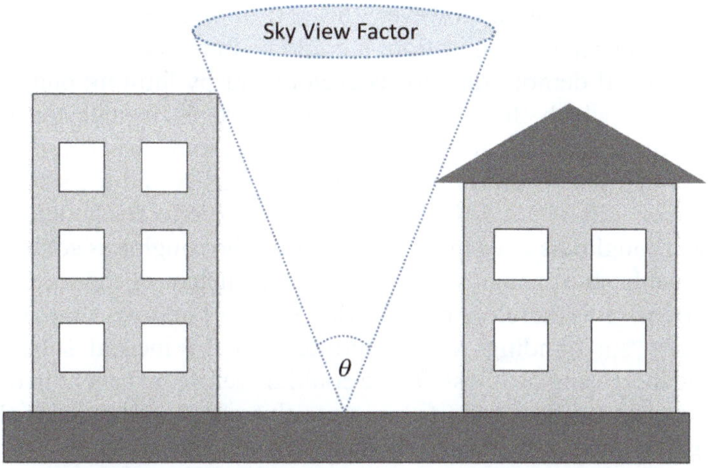

Figure 4.28. Schematic of the sky view factor principle.

Figure 4.29. Skyview image of 5th Avenue, Manhattan, New York taken with a fisheye lens (This fisheye image has been obtained by the author from the Wikimedia website where it was made available under a CC BY-SA 3.0 licence. It is included within this article on that basis. It is attributed to Autopilot.).

behind, and tall buildings beyond, which then decrease in size again towards the next area of green space. This effectively maximises the sky view factor for all locations and allows maximum penetration of cool air from the park into the surrounding urban context. The aim of this urban planning policy was to allow all buildings to be able to 'see the green spaces' and to be able to effectively shed heat as thermal radiation. In Ho Chi Minh this was implemented across a large area of the city that was undergoing rapid growth. However, this sort of intervention is not only limited to areas undergoing rapid urbanisation; it could be implemented in existing cities as they undergo natural demolition and redevelopment by limiting building heights.

Another way in which the UHI can be mitigated is through the use of urban planning to increase surface roughness. If we recall from earlier in the chapter, the boundary layer depth δ depends upon the upwind fetch x and the surface roughness over a distance of $\sim$20–100 $\times$ δ. The boundary layer is effectively determined by the average surface roughness over the fetch, and that the roughness sublayer gives way to the inertial sublayer at roughly twice the average height of the roughness features (buildings or trees). It is therefore possible that trees or buildings that are significantly larger than their surroundings will penetrate into the inertial sublayer, creating turbulence and increasing heat loss. There could therefore be benefit in dispersing tall buildings throughout urban areas to promote this effect, rather than clumping tall buildings together and enhancing the heat island. For example, the 'Shard' in London is surrounded by smaller buildings and will likely penetrate the inertial sublayer, increasing turbulence, dispersing air pollution, and increasing convective heat loss. In contrast, the Eastern Cluster of the city of London (see figure 1) on the opposite side of the river Thames, contains several tall buildings in close proximity to each other, which will enhance the UHI and lead to increased air pollution and temperatures.

The magnitude of the UHI for larger cities is equivalent in magnitude to all but the most aggressive projections of future climate change. If the UHI can be reduced through the incorporation of green and blue infrastructure or through urban planning policies, then it may be possible to offset some of the impacts of climate change. The most suitable intervention will depend upon not only the background climate of the city but also what other ecosystem services or public amenities may be of most benefit to the populous. Since climate change will alter the patterns of rainfall over the year, leading to prolonged periods of warm weather with little rainfall, and drought-resistant plants have reduced evapotranspirative cooling potential, it is best to incorporate green infrastructure as part of a water retention SuDS scheme so that evapotranspirative cooling can be provided in the summer months when it is most needed. In this way, urban planning links the issues of urban water management and UHI reduction, allowing common solutions to be implemented in urban areas in the form of green and blue infrastructure.

References

[1] Gunawardena K R and Kershaw T 2017 Urban Climate Influence on Building Energy Use in *Proceedings of International Conference on Urban Comfort and Environmental Quality* (Genova, Italy: Genova University Press) September 2017

[2] Gunawardena K R, Wells M J and Kershaw T 2017 Utilising green and bluespace to mitigate urban heat island intensity *Sci. Total Environ.* **584** 104–1055

Chapter 5

Weather extremes

Climate is a difficult thing for people to comprehend because we tend to think in terms of the daily and seasonal variations we call weather. Extreme events such as heavy rainfall causing floods, snowfall causing travel disruptions, or heatwaves stick in our minds. 2003 was not a particularly warm year, in fact the summertime temperatures were about average, but the two week heatwave that sat over Europe in August, which claimed ~70 000 lives, biases our opinion about the summer of 2003. Such events are extreme not only in that they occur infrequently but also that they have profound impact on people and urban areas. As the climate changes as a result of human activity and anthropogenic carbon emissions, the relative probabilities and likelihoods of extreme events will change. The intensity of such events will also change, with hotter, longer heatwaves, and more powerful storms, with higher wind speeds and more rainfall; therefore, what is currently classed as extreme weather may become the norm in the future. Unfortunately, the low-probability nature of extreme weather means that there are few events in historical records, which makes predicting future re-occurrences difficult. As such, there is a lot of uncertainty in climate change projections when it comes to lower probability events such as powerful storms and heatwaves. However, due to their destructive nature to property, infrastructure, and human lives, there is much ongoing research to improve the future forecasting of such events.

5.1 Heatwaves

Heatwaves are periods of calm hot weather (for that location) that persist for several days; this may also be accompanied by high humidity in marine climates. There are several different definitions of what constitutes a heatwave depending upon the region, a common definition which has been used for the UK is three or more consecutive days with daily maximum temperatures above 28 °C, not dropping below 16 °C at night. This definition is useful as it highlights the risk to human health in that the human body is able to withstand high temperatures for short

doi:10.1088/978-0-7503-1197-7ch5

periods, so long as it is able to shed stored heat when temperatures drop, usually at night. Another common and more international definition of a heatwave is the one used by the World Meteorological Organisation: five or more consecutive days where the daily maximum temperature exceeds the average maximum temperature by 5 °C; the baseline period being 1961–90. This definition is useful because it recognises that heatwaves may not always be in the height of summer, and that a particularly hot spring or autumn can be just as dangerous as higher temperatures in summer, due to the time lag associated with the adaptive response of human physiology. However, it does ignore the dangers associated with higher daily minimum temperatures and the associated implications for human health and wellbeing.

Estimates of the number of deaths across Europe as a result for the 2003 heatwave vary from ~15 000 up to ~70 000, depending on the methodology used and the assumptions made regarding how many people would have died during that period or in the following few months. Regardless of the methodology used to estimate the death toll, heatwaves have a profound impact on the populous, especially on vulnerable groups such as the elderly, infirm, or very young. The deaths that occurred as a result of the 2003 heatwave mainly occurred in large urban areas with a distinct UHI, which itself is intensified by the atmospheric conditions typical of a heatwave. The August 2003 heatwave is thought to be the warmest period for up to 500 years, and many European countries experienced their highest temperatures on record; in the UK temperatures peaked at 38.5 °C, and the London UHI was measured to be 9 °C. As we have discussed in previous chapters, high temperatures and reduced thermal comfort have a negative impact on worker productivity. The 2003 heatwave is estimated to have cost the UK economy between £400–500m in reduced manufacturing output.

Global populations are ageing with people living longer in both developed and developing countries. The UK is no different. The percentage of the UK population aged 75 or over is expected to increase from ~8% in 2015 to ~18% by 2085. When temperatures exceed 35 °C, all age groups are at risk; however, it is the elderly (and other vulnerable groups) who are most at risk and likely to be adversely affected by extremes of temperature. It is entirely likely that humans will undergo some form of autonomous physiological acclimatisation to changes in mean temperatures over time. However, research suggests that humans are less able to adapt to rapid, sudden changes in temperatures, particularly if the overall annual temperature variability increases as it is expected to as a result of climate change.

Cold is still the main cause of temperature-related deaths in the UK, with between 35 800 and 49 700 cold-related deaths per year on average. As the climate warms, cold will remain an important climate risk even though winters will become milder overall because minimum temperatures will remain similar to those we experience now. This coupled with the generally poor thermal performance of the UK housing stock and a growing, ageing population means that the number of cold-related deaths is unlikely to change much in the near future. Current estimates indicate only a ~2% reduction in cold-related deaths in the UK by 2050 under a medium emissions scenario (A1B). Current measures to improve the energy efficiency of homes will also reduce the impact of cold weather on the occupants. However, increasing the airtightness of the building fabric and increasing the level of insulation has to be undertaken carefully so as not to increase the risk of overheating, and provisions must be made to allow the removal of heat in the summer months.

The number of hot-days per year has been increasing steadily since the 1960s, and currently there are around 2000 heat-related deaths in the UK each year. As the number of hot days per year is increasing as a result of climate change, so is the risk of a severe heatwave. All but the lowest estimates of climate change (RCP2.6) show that by the 2040s a summer as hot as 2003 will be very common and will considered cold by the end of the century [1]. The ageing population means that the number of vulnerable people is increasing, and thus the number of premature heat-related deaths is also expected to increase. The number of heat-related deaths in the UK is predicted to more than triple by the 2050s, and increase five-fold by the 2080s, under a medium emissions scenario (A1B, median estimate) from the baseline of 2000 deaths per year. This could be further exacerbated by poorly implemented energy saving measures such as increased airtightness and insulation unless also accompanied by increased ventilation provision for example. There is evidence that newer homes are at a greater risk of overheating than older designs, due to improved airtightness and greater insulation, but also due to the removal or masking of thermal mass by use of modern construction materials and methods. At present there are no comprehensive policies in place to adapt the existing building stock to higher temperatures, reduce the urban heat island, or to safeguard homes against heatwaves. There is evidence that most people lack a basic understanding of the risks to health from elevated indoor temperatures, and therefore are unlikely to take measures to safeguard their and their dependents' wellbeing. Particular care will therefore need to be applied to buildings such as schools, hospitals, care homes, and prisons.

Heatwaves by definition are not a common occurrence. In addition to August 2003, other recent heatwaves across the UK were July 2006, July 2010, October 2011, March 2012, July 2013, July 2015, September 2016, and most recently June 2017. While it may appear that heatwaves are becoming more common (which they are due to a warming and more variable climate), these heatwaves were not as disastrous as the 2003 heatwave. This is likely due to the preceding 2003 summer months being abnormally cool, with sudden changes in temperature being more dangerous to humans. Ongoing research into the return periods of extreme weather events such as heatwaves has shown that human activities have increased the likelihood of heatwaves considerably; an event that would have been expected to occur twice a century in the early 2000s can now be expected twice a decade. Furthermore, a more severe 2003 type event has increased from 1 in several thousand to ~1:100 in little over a decade, and is expected to be an average summer by 2040 [1].

As the climate warms, the distribution of summer temperatures is expected to shift to higher warmer regimes, but also stretch, increasing overall variability in temperature, which would further increase the frequency of heatwave events (using the current definitions). Estimates of the effects of climate change on heatwave events are varied. For a 2 °C increase in global mean temperatures, it is estimated that a heatwave event in Europe may be expected to increase in intensity by between 1.4 °C and 7.5 °C [1].

The methods we can employ to keep our buildings cool and safeguard ourselves against high temperatures are similar to those already covered in chapters 3 and 4. Due to the characteristics of a heatwave (high temperatures, intense solar radiation, and high atmospheric stability with little wind), ventilation may be inefficient or even harmful unless the air can be cooled somehow, by evapotranspiration from plants for example, since the external air may be hotter than that inside the building. Due to reduced wind speeds, cross ventilation may well be ineffective at removing heat. Furthermore, thermal mass, which is effective at reducing peak temperatures, may well contribute to higher mean internal temperatures during a prolonged heatwave, particularly if exposed to direct sunlight, unless stored heat can be purged at night. Effort is best placed at keeping heat out of our buildings, such as closing curtains, blinds, or shutters early in the day to keep solar radiation out of the building. This is particularly important for west-facing rooms, which get afternoon and early evening sun, and are the most prone to overheating. External shading also will be effective, whether in the form of roof overhangs, balconies, brise soleil, or even just trees and vegetation; these will reduce the ingress of solar radiation and reduce internal temperatures. During a heatwave where nocturnal temperatures do not drop significantly at night, building heat loss via radiation and convection will be reduced at night; therefore, windows should be left open to allow heat to escape from the structure, purging any thermal mass, which would be enhanced due to the pressure difference between warmer air inside the building compared to outside. If possible, openings on multiple storeys should be used to also utilise the change in pressure with altitude (so-called stack ventilation); however, ground floor openings can come with a security risk. Furthermore, it is imperative that these windows or openings are closed again in the morning, before external temperatures increase, to retain the coolth within the building. Changing the colour of the building will help reduce the risk of overheating by reflecting solar radiation, and additionally increasing the amount of insulation in lofts will assist in keeping homes cool since the majority of solar heat gains are via the roof space.

Finally, more drastic measures to reduce overheating may focus on adding or increasing access to thermal mass within the building structure (coupled with effective night-time ventilation). Modern building materials such as plasterboard are effective insulators, and a dry-lined wall effectively masks any thermal mass behind (the same is true of suspended ceiling tiles). Replacing plasterboard with wet gypsum plaster allows heat to be absorbed from the air by the bricks/blockwork behind. Similarly, replacing carpets or wooden flooring with stone flagstones or ceramic tiles will allow heat to be stored in the screed/concrete floor. These effectively reduce not only the air temperature but also the mean radiant temperature since the surfaces around us are at a lower temperature; this has the effect of making us feel cooler because our bodies, which are almost perfect receivers/absorbers of thermal radiation (a black body), are receiving less radiation from our surroundings, and hence are able to radiate heat more effectively. Ultimately, any heat stored within the building structure needs to be lost to the environment eventually. Either the building needs to be made so thermally massive that the building will take several weeks to reach equilibrium with the elevated external temperatures (e.g. complete stone/concrete construction like a church) or else ventilation must purge heat at night when temperatures are lower. Increasing the ventilation provision of a building by adding a secure floor and ceiling-level vents (preferably on opposite façades) will allow stack-driven ventilation to occur at night, thus removing heat from the building and purging the thermal mass for a new day.

5.2 Storms

While I write this in 2017, storm Aileen is battering my house with 100+ km h^{-1} winds and torrential rain. Meanwhile, across the Atlantic hurricane Irma is causing wide-spread devastation to Florida after already passing through the Caribbean and the Bahamas. The severe impacts of high winds, torrential rain, and storm surges seem highly evident. The number of severe storms per decade has been rising steadily since the 1950s, with the stormiest years coinciding with a sustained positive North Atlantic Oscillation (NAO) index, a measure of sea level pressure anomaly versus the long term average. A positive value of NAO indicates higher pressure and is typical of stronger westerlies over the mid-latitudes, more intense weather systems, and wetter and milder weather over western Europe. Whereas a negative NAO blocks storms moving up from the Gulf of Mexico towards the UK, and the least stormy years are associated with more negative NAO pressures. It seems evident that there is an association between the NAO and the El Niño/La Niña Southern temperature oscillations over the Pacific, but this relationship is not well understood due to the many other factors involved, such as sea ice melt in the Arctic. Unfortunately, this makes predicting future storms and wind speeds very difficult, particularly when considered alongside the other uncertainties associated with estimates of climate change. At the time of writing, the UK Met Office has insufficient confidence in estimates of wind speed from its climate models to include the outputs in the publically available UKCP09 climate projections. This arises largely from the fact that UKCP09 is based upon many different climate models, not just those from the UK Met Office but also those from other Metrological Institutes worldwide. This allowed the formation of the probabilistic climate change projections that have been used throughout this book; however, for wind speed and storminess the results from the various models were so divergent that no confident consensus could be derived. This means that this section is based upon less evidence than the previous sections of this book; however, the risks to buildings, infrastructure, and human life require that it is given consideration. Furthermore, it seems logical that given the trends of increased winter rainfall and increased temperatures that the number and intensity of storms will increase.

People still talk of the 'Great Storm of 1987' that devastated the south of England. Schools were evacuated and pupils sent home (myself included), and people were advised to stay at home where it was safe; meanwhile, outside trees were uprooted, fences and walls were blown over, and roofs were ripped off buildings. In the coastal village where I lived, sea defences were destroyed and sections of coastal roads and footpaths washed away. This storm is considered to herald the onset of increased storminess, with the 1990s being the UK's stormiest decade since the 1920s; however, the 2010s seem set to take this title.

The UK experienced a spell of extreme weather during the winter of 2013/14 (figure 5.1). Between late January and mid-February 2014, a succession of major storms brought flooding and destruction to the UK. During these few weeks around six major storms separated by two to three days hit the UK; these followed an earlier stormy period during December and early January. Overall, during this winter period (December–February) at least 12 major winter storms hit the UK, making this the stormiest period for at least 20 years. January 2014 was the wettest January on record, with more than five months' worth of rainfall recorded in some locations. The most severe of this rapid succession of storms occurred in early-mid February, with the ground still saturated and inundated from rainfall over the new-year period.

These storms resulted in numerous weather-related impacts across most of the UK throughout the winter period. There was major flooding, with the Environment Agency reporting 6000+ properties flooded. The Somerset Levels were very badly affected, with large areas remaining under water from late December through the entire winter period to mid-March. There was also severe flooding along much of the River Thames through Oxfordshire, Berkshire, and Surrey. In addition to flooded properties and businesses, transport infrastructure was also affected, with many roads underwater and several villages on the Somerset Levels being only accessible by boat. The Southwest mainline railway between Exeter and Bristol was also affected by flooding on the Somerset Levels, and was shut down until mid-March. The following online video clip from the Met Office shows the extent of some of the impacts that these storms caused.

Figure 5.1. Weather summary video clip of winter 2013/14 (Crown Copyright. Contains public sector information licensed under the Open Government Licence v1.0). Available from https://doi.org/10.1088/978-0-7503-1197-7.

Figure 5.2. Huge waves batter the Cornish village of Porthleven on 5th February 2014 (©Matt Clark, Met Office).

As well as inland flooding, strong winds and huge waves battered exposed coastlines in the South and Southwest, causing widespread destruction to property and infrastructure. Strong winds, high tides, and tidal surges acted in combination to produce huge waves to batter the coastline (see figure 5.2). The swells produced by the storms had a long wavelength, allowing for the formation of waves with a large amount of speed and energy, and able to reach record heights. The most severe storm during this winter period was on 12th February, with waves of 25 m (82′) recorded off Southern Ireland and wind speeds reaching hurricane force (see table 5.1).

As we can see from figure 5.2, coastal conditions can become exceptionally dangerous. This is not only a danger to human life but can also cause considerable damage to buildings and infrastructure as huge waves overtop coastal defences. The mainline railway through Devon to Cornwall was severely damaged along a coastal stretch of line at Dawlish during the storm of 4–5th February (see figures 5.3 and 5.4), severing a key transport link to the Southwest peninsula for several weeks. This sort of disruption has a knock-on effect for the whole of the rail transport network since many national rail services begin or end in the Southwest along the stretch of line beyond Dawlish. This meant that a significant proportion of the rolling stock became trapped on the Southwest peninsula, causing widespread disruption across the entire rail network. These storms caused considerable coastal erosion, and the rapid succession of storms exacerbated coastal damage since weakened defences could not be repaired before the next storm hit.

Figure 5.4 shows damage to the substantial retaining wall supporting the foundations to the mainline railway. This damage caused by waves overtopping coastal defences is likely a result of the succession of storms progressively weakening the wall. This highlights the need for a more rapid response after storms to check and repair sea defences and infrastructure. It would seem probable that as the climate warms and contains more energy, the weather will become turbulent,

Table 5.1. The Beaufort scale used to describe wind intensity.

Beaufort wind scale	Wind description	Wind speeds in mph (km h^{-1})	Sea description	Probable wave height (max probable) (m)
0	Calm	<1 (<1)	Calm (glassy)	0 (0)
1	Light air	1–3 (1–5)	Calm (rippled)	0.1 (0.1)
2	Light breeze	4–7 (6–11)	Smooth (wavelets)	0.2 (0.3)
3	Gentle breeze	8–12 (12–19)	Slight	0.6 (1)
4	Moderate breeze	13–18 (20–28)	Slight – Moderate	1 (1.5)
5	Fresh breeze	19–24 (29–38)	Moderate	2 (2.5)
6	Strong breeze	25–31 (39–49)	Rough	3 (4)
7	Near gale	32–38 (50–61)	Rough – Very rough	4 (5.5)
8	Gale	39–46 (62–74)	Very rough – High	5.5 (7.5)
9	Strong gale	47–54 (75–88)	High	7 (10)
10	Storm	55–63 (89–102)	Very high	9 (12.5)
11	Violent storm	64–72 (103–117)	Very high	11.5 (16)
12	Hurricane	73+ (118+)	Phenomenal	14+ (–)

Figure 5.3. Damage to the Southwest mainline railway at Dawlish (Devon) after the storm of 4–5th February. Breach of sea defences and destruction of retaining wall, left rails suspended in mid-air after foundations were washed away (©Matt Clark, Met Office).

Figure 5.4. Photo showing damage to the retaining wall bounding the mainline railway at Dawlish (©Matt Clark, Met Office).

resulting in more and more powerful storms. A lack of preparedness will only exacerbate the effects of extreme weather.

The winter of 2015/16 was also exceptional; the mildest December on record was also the wettest, and this was coupled with the seemingly continuous series of storms that swept across the UK. In total, 509.2 mm (~20 inches) of rain fell in the three-month period December 2015 to February 2016. December 2015 was the warmest December since records began in 1910, with mean temperatures about 4 °C above the long-term average. The UK Met Office states that there is a direct link between the warmth and the record levels rainfall, with that December being also the wettest of any calendar month of record, with almost double the normal amount of rainfall at 218.8 mm.

In 2015 the UK Met Office began naming storms rather than just hurricanes, with the first named storm being Abigail which hit UK shores on 10th November 2015; over the next few months there was a rapid succession of storms (see table 5.2) that caused disruption and flooding. The three storms in December provided much of the total month's rainfall, with the storms propelled by the Jet Stream towards the UK, and additional contributions from the El Niño weather phenomenon and anthropogenic climate change. The Jet Stream, which flows from the Gulf of Mexico towards the UK, is responsible for much of our mild weather, and is indicative of a positive NOA. In this case, the positive NOA and strong Jet Stream will have contributed towards the uncommonly mild weather, with the elevated temperatures resulting in greater evaporation of water and leading to increased rainfall.

The 2016/17 storm season saw only five storms, but as the 2017/18 storm season (September–March) begins with storm Aileen currently raging outside, we can only wonder what the future holds.

Table 5.2. Names and dates of storms in the 2015/16 storm season.

Name	Date of impact on UK and/or Ireland
Abigail	12–13 November 2015
Barney	17–18 November 2015
Clodagh (Clo-da)	29 November 2015
Desmond	5–6 December 2015
Eva	24 December 2015
Frank	29–30 December 2015
Gertrude	29 January 2016
Henry	1–2 February 2016
Imogen	8 February 2016
Jake	2 March 2016
Katie	27–28 March 2016

Building codes in the UK already take account of the impact of wind shear forces on buildings. However, projections of climate change indicate that storms will become more powerful, meaning that additional care will need to be applied to the design of buildings in exposed areas where squalls of very strong winds are common, even if they are typically only of a short duration. Violent storms have a variety of names according to the region in which they occur; hurricanes, typhoons, and cyclones are all basically the same phenomenon, consisting of high-speed ~120 km h^{-1} (~75 mph) rotating winds. This is mean wind speed, and individual gusts can be significantly stronger and more powerful. These storms can cause extensive damage (table 5.3), not only by the direct effects of the powerful winds and intense, torrential rainfall, but also indirectly by associated flooding, storm surge, and landslides.

The energy contained within wind and its destructive power increases with the square of its velocity; hence, a 100 km h^{-1} gust can exert a force 100 × greater than a 10 km h^{-1} gust. To resist the lateral forces of these winds, buildings are typically made more rigid or braced somehow. However, more primitive architectures in tropical climates adopt the philosophy of flexibility, allowing structures to sway and give in the winds like the palm trees of such climates. The alternative to these solutions is to shelter the building either via wind breaks or by reducing the profile of the building, making it more aerodynamic or sinking it into the ground.

A building in a location prone to cyclonic storms must be able to withstand the extreme winds that can blow from every direction as the storm moves overhead. The building must also be able to withstand the rapid and extreme changes in air pressure generated by the storm. Severe storms are characterised by this rapid change in pressure, differing them from severe gales. This change in pressure is >±10 kPa (>±10 mb) in a three-hour period. This coupled with the local changes in pressure due to wind flow and the Bernoulli principle (cf. the lift produced by plane wings due to the pressure difference between the top and bottom) means that the pressure across different building elements can be extreme. Additionally, as we have previously discussed, the drop-in pressure due to an intense cyclonic storm leads

Table 5.3. Wind gust velocity and damage to vegetation and property.

Wind gust velocity	Damage
75 mph–93 mph 120 km h^{-1}–150 km h^{-1}	Shrubs, trees, and foliage but unlikely for buildings in good repair.
94 mph–112 mph 151 km h^{-1}–180 km h^{-1}	Considerable damage to shrubs and tree foliage, with some trees blown down. Some damage to building roofs, windows, and doors; minimal damage to building structures.
112 mph–130 mph 181 km h^{-1}–210 km h^{-1}	Extensive damage to trees and shrubs, with large trees blow down. Some damage to building roofs, windows, and doors; minor damage to curtain-walls and small structures.
131 mph–149 mph 211 km h^{-1}–240 km h^{-1}	Shrubs and trees blown down. Extensive damage to roofs, windows, doors, and some curtain-walls. Complete failure of roofing structures on small buildings.
150 mph + 241 km h^{-1} +	Shrubs and trees blown down. Considerable damage to buildings; roofs, windows, and doors. Complete failure of roofing structures on small buildings. Curtain-wall and façade failure of industrial type buildings. Small buildings may be overturned completely.

to a rising sea levels, which can lead to coastal or estuary flooding, particularly if the storm coincides with an astronomical high tide.

Roofs are particularly prone to being damaged or destroyed by the high suction (negative pressure difference) on their leeward side (see figures 5.5 and 5.6) due to the rapid flow of air over the building or a combination of suction on top and pressure underneath the roof overhangs. In taller buildings, the influence of wind gusts and the pressure differences between the windward and leeward sides of a building can set up oscillations in the building structure, which can result in structural failure.

The centre of a cyclonic storm is at extremely low pressure, and if this hits a building that is particularly well-sealed, too quickly, the normal pressure (or even elevated pressure due to temperature differences) inside the building can cause it to explode outward and collapse. This is perhaps an extreme example, but the failure of window and door seals leading to increased air leakage (infiltration) and potentially increased energy bills in the longer term if not remedied is much more likely. This presents a problem as the current tendency is to create more insulated and air-tight buildings on the pretext that they are more energy efficient.

Buildings can be protected from high winds by planting trees, but care should be taken not to plant them closer than ~15 m in case the tree is blown over. The sheltering effect of such vegetation will be effective only over a distance of ~8–10 times the height of the tree. In undulating or hilly areas, building construction along ridges should be avoided since they will experience accentuated wind velocities, whereas lower down in a valley will typically experience lower speeds. However, the reverse may be true in long narrow valleys where wind becomes funnelled along the valley floor; this obviously depends on the orientation of the valley to prevailing winds.

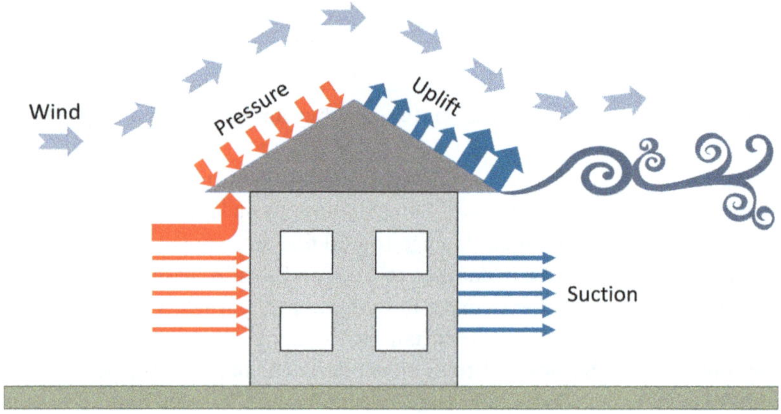

Figure 5.5. Illustration of the forces on a building caused by wind pressure and turbulence.

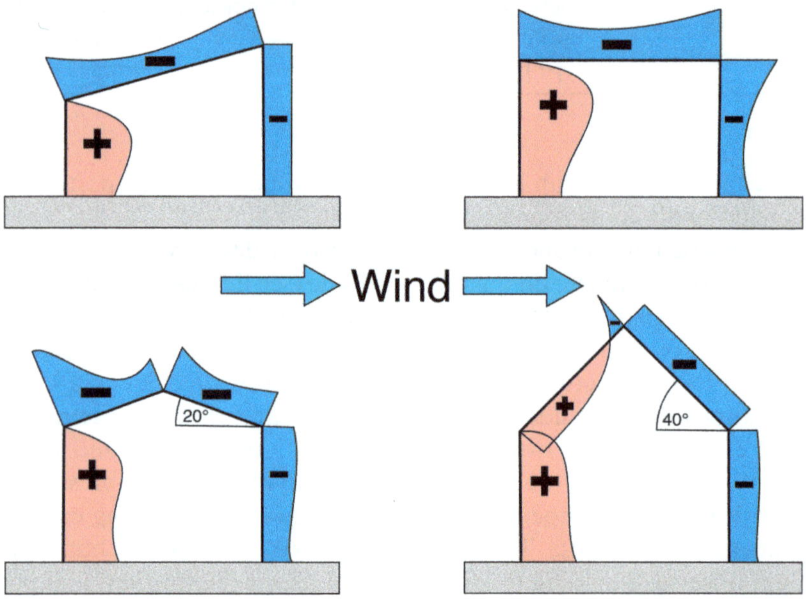

Figure 5.6. Forces exerted on different roofs and façades by the wind; positive signs indicate pressure, while negative signs indicate suction.

The shape of a building is the single most important factor in determining the performance of buildings in cyclonic storms. Compact, simple geometric shapes are best. Buildings should ideally be symmetrical, with square, or even hexagonal floor plans, and roofs should have multiple slopes because these present less area to winds and will perform better under wind load forcing than gable roofs with two slopes. If the site dictates a rectangular shape, then it is best if the length is not more than 3 × the width. Experiments have shown that hip roofs with a pitch in the range 25°–40° are most resistant to high winds. Meanwhile, roofs with a pitch <22° should be avoided due to the increased suction from wind flow (figure 5.6).

All roof elements, such as rafters and purlins, should be securely fixed to the main structure. Additionally, the double nailing of roof tiles is advisable to prevent removal by suction forces. Roof canopies and overhangs should be minimised or avoided if possible, although herein lies a contradiction with reducing temperatures by shading of façades. In some climates, brackets fixed to the main building structure or to the ground can be used to tie down the roof before the storm season starts. Verandas or covered walkways should be built as separate structures rather than extensions to the main structure so they can be blow off without damaging the main structure. Windows and doors are a particular danger in high winds; if these are broken by flying debris or breached by changing pressure levels, wind will have easy access to the building interior, and the resultant uplift pressure in the roof space may cause the destruction of the entire building structure. Furthermore, frames should be securely anchored to the rest of the building structure, and large expanses of glass should utilise shatterproof, toughened glass, and be protected by storm shutters. Many of these design features can be seen employed in the traditional buildings of the world's exposed locations.

Reference

[1] Christidis N, Jones G S and Stott P A 2014 Dramatically increasing chance of extremely hot summers since the 2003 European heatwave *Nat. Clim. Change* **5** 46–50

Chapter 6

Conclusions

In this book, we have seen that the global climate is changing, with anthropogenic greenhouse gas emissions the dominant cause. The global increase in temperature of ~1 °C since 1880 (see figure 1.15) is also mirrored in the UK climate, with higher average temperatures (figure 1.13). Average temperatures across the UK have increased in line with global projections of climate change, with a trend towards milder, wetter winters and warmer, drier summers. Sea levels globally and around the UK have risen by between 15–20 cm since 1900, and there is a growing amount of evidence that climate change is partly responsible for the severity and frequency of recent extreme weather events.

With very few exceptions, the building industry has yet to face up to the implications of climate change and the effect this will have on the performance of our buildings, not just in terms of energy usage but also in terms of thermal comfort and the effect this has on the productivity, health, and well-being of the building occupants. The building industry is still coming to terms with how to reduce the impact of our buildings on the climate via reducing carbon emissions and improving efficiency as part of the mitigation agenda, let alone started to consider how to reduce the long-term impact of the climate on our buildings. The buildings we design now will still be in use when the impacts of climate change make themselves felt, and in the case of residential buildings, many of the homes we will be occupying at the end of this century have already been built, many before the introduction of building regulations (1965). One important realisation is the fact that the weather data we base the design of our building on is out of date, being based upon the averages of historical observations, primarily from the previous century. Despite the recent efforts of academics and professional bodies such as the Chartered Institution of Building Services Engineers (CIBSE) to update this weather data, such historical observations will inevitably lag behind the changing climate. Therefore, it is becoming increasingly important to assess how buildings will perform in the future not just based on the current or historic climate.

doi:10.1088/978-0-7503-1197-7ch6

The recent Paris Agreement (2015) represents a significant step forward to curb carbon emissions and limit future climate change. A total of 195 nations, including the UK, will 'pursue efforts' to limit the global temperature increase to <1.5 °C. For this ambitious goal to be achieved, global carbon emissions will need to peak soon and decline rapidly afterwards. Currently, the commitments of partner countries will still lead an estimated 2.7 °C rise in global temperatures, indicating how much further there is still to go. The uncertainties associated with climate change predictions (see chapter 1) mean that even if the Paris Agreement targets are met there is still the possibility of a ~4 °C warming of global temperatures by the end of the century. Furthermore, we need to consider that these are global averages, and that by 2100 there will still be significant lag associated with the thermal inertia of the oceans and thus land surface temperature increases will be significantly higher than 1.5 °C. We therefore need to consider how we will cope with increasing temperatures of several °C and the implications this will have on the weather and performance of our buildings, without compromising the ongoing mitigation agenda. The knowledge contained within this book is not new, and much of it has been employed in various locations around the world for millennia. In order to increase the resilience of our buildings and urban areas, we simply need to apply it.

6.1 Building resilience

This book has considered how buildings in different locations have evolved to become resilient to the local climate, whether that climate is warm and humid, hot and dry, temperate, marine, or arid. By examining the responses of different locations to the environmental stimuli we can consider how to adapt our building design to cope with anticipated future climates.

The way we use our buildings in the UK (our culture) is very different to that of the hotter Middle Eastern or even Mediterranean countries. Elsewhere, buildings are shut up during the day to keep the heat out. Here, we like large windows and openings to allow light into our buildings and to provide views of the surrounding landscape, and we are more concerned about heat loss in winter than heat gains in summer. The introduction of double glazing, originally to provide noise reduction, is now standard due to its status as an energy saving feature. This, along with improved airtightness and increased insulation levels are some of the reasons why we are now seeing instances of overheating in new buildings. Each pane of glass reduces the amount of light passing through a window by ~10% but glass being opaque to thermal infrared radiation stops ~50% of heat getting out again, a value which is often improved further through the coating of the glass with a thin layer of metal, so-called low-ε (emissivity) glass. This is why in Nordic countries you will see triple and even quadruple glazing. You might have expected that the widespread adoption of double glazing would lead to a change in the way we design our buildings, but this does not seem to be the case. Of course this extra heat retention is a benefit in winter, but it can be a hindrance during warmer sunny weather, particularly when the Sun is lower in the sky (early morning or late afternoon, or during the shoulder months in spring and autumn) since solar radiation can penetrate farther into a room, heating surfaces and reducing the likelihood that radiation (which is radiated in every direction) can get out of the building again. As the climate warms, design features such as double glazing, which are intended to keep heat in in winter, may become a major cause of overheating unless steps are also taken to keep heat out in summer. Shutters over windows such as one sees in the warmer Mediterranean climates (see figures 3.16 and 3.18) would seem a logical solution; however, this would be a serious departure from the British culture. The challenge therefore is how to design new buildings and adapt existing ones to this changing climate without fundamentally changing the way we use our buildings.

There are several examples of successful attempts to produce climate change resilient buildings that combine aspects of both mitigation and adaption, both in domestic and non-domestic buildings. The recent 'Design for Future Climate' tranche of industrial research projects funded by the then Technology Strategy Board saw ~50 real building projects across the UK consider how to adapt to a changing climate. The projects ranged from eco-towns, to hospitals and laboratories, and from commercial offices to swimming pools. Sadly, although plans of adaptions to the building design had to be drawn up and timelines of any future interventions provided as a stipulation of the project funding, actual incorporation of these interventions into the constructed building was not required.

There are, however, several built examples of how to make buildings resilient to climate change while at the same time reducing energy usage to satisfy the mitigation agenda. One such example is Montgomery Primary School located in Exeter, which embodies many of the principles we have discussed throughout this book. As part of the UK Governments 'Building Schools for the Future' incentive, several schools around Exeter and the Southwest were retrofitted, and others demolished and rebuilt. However, this exercise was not overly successful, with several schools reporting overheating issues, requiring remedial works and leading to well-publicised lawsuits between the local authority and contractors. Montgomery School was designed in response to this, with the school designed to still pass current overheating criteria in the 2050s. The initial brief was also for the school not to have a heating system, with heat provided by the pupils and stored overnight and over the weekends. This was later relaxed, but the school was still designed to be zero-carbon in use. To achieve this, the school was designed according to the stringent Passivhaus standard, and features triple glazing, a highly insulated and airtight building envelope, and a mechanical ventilation system with heat recovery to extract heat from exhaust air to warm incoming fresh air. Montgomery Primary School was the first certified Passivhaus school in the UK, and has also been shown to be zero-carbon in use (over the course of a year) thanks to the large array of photovoltaics (figure 6.1) on the roof. It was awarded Building of the Year by CIBSE in 2014. While this is all very impressive, we are concerned here with its climate change resilience credentials.

Many Passivhaus buildings are lightweight in construction, not because the standard stipulates this, but because the super-insulated façades tend to be wooden frames filled with insulation (figure 6.2) to create very responsive buildings that can be heated or cooled easily. In order to store metabolic heat from pupils, we require a

Figure 6.1. View of the Southern side of Montgomery Primary School in Exeter, showing the large array of photovoltaics and its shading effects. Photo taken midday 16th October 2012.

Figure 6.2. Cross-sections of a typical super-insulated wall build up (left), and the heavyweight walls used at Montgomery (right). Both contain 150 mm (6″) of insulation.

building with significant thermal mass. The entire school is made from concrete, with 150 mm of expanded polystyrene insulation sandwiched between 150 mm (6″) of concrete on the inside and 75 mm (3″) of concrete on the outside surface (figure 6.2); the ceilings and floors are also concrete. The concrete is fair faced (smooth) and is not plastered, with only paint applied on the walls for aesthetic reasons and to improve the daylight uniformity in the classrooms.

All this concrete not only provides the building with a huge amount of thermal mass to regulate internal temperatures and to store metabolic heat, solar gains, and other internal heat gains from lights and IT for example, but it also has the benefit that cast concrete such as this is sufficiently airtight to meet the Passivhaus standard. Meanwhile, brick, block, or wooden frame constructions (such as in figure 6.2) will require an additional air-tight layer, either a membrane embedded in the construction or gypsum plasterwork on the interior surface. Concrete is also resistant to wind shear forces, driving rain penetration, and flood damage. Montgomery Primary School is located close to the river Exe, in an area that has historically been prone to flooding. For this reason, the new school building was raised 1 m above the level of the existing school, a tactic now employed in the Netherlands after the removal of dikes as part of the 'Room for the River' project.

Schools in the UK are some of the most stringently designed buildings, with guidance and regulations pertaining to air temperatures, CO_2 concentrations, ventilation rates, acoustics, and daylighting. This poses several problems as in order to meet the daylighting requirements windows have to be large, but if exposed to direct solar radiation then can easily lead to overheating. Additionally, a standard 8×7 m classroom, such as those at Montgomery, will contain ~30 pupils, at least one teacher, lighting, and IT (e.g. PC and smartboard). This can easily equate to ~5 kW of heat being generated inside the room, which will need to be regulated by the structure or removed by ventilation. It is for this reason that schools and offices are prone to sudden overheating when the Sun comes out. It was decided early on in the design process for all the classrooms to be located on the north façade, which is two storeys, and offices and staff rooms to be located on the south façade. North-facing windows in the Northern Hemisphere will only receive diffuse sunlight, and thus are ideal for daylighting purposes as there is limited risk of glare or overheating. Comparison of figures 6.1 and 6.3 shows that the windows on the southern façade are smaller and are shaded by either brise soleil or from the overhanging roof and photovoltaics. The photos shown here were taken during the official opening of the school on 16th October 2012, and hence the roof shading is not fully covering the south-facing windows in the images but would do so in the late spring and summer months. This design of the shading device allows passive solar heating of the building in the winter months when solar gains are beneficial, but prevents direct solar radiation entering the building in the summer. Internal blinds (see figure 6.1) are provided to limit solar gains in the shoulder months if required, and to reduce glare.

The activities within schools generate a large amount of heat, but they are also used primarily during the UK heating season. Therefore, controlling the flow of heat within the building is key not only to reducing heating bills but also to preventing overheating, both now and in the future. The high thermal mass structure of the building utilises fair faced concrete walls, a polished concrete floor (containing crushed glass to improve the aesthetics), and a concrete ceiling. All these hard parallel surfaces mean that measures need to be deployed to control the acoustics and reduce the reverberation time in order for the school to be a viable learning environment. This normally takes the form of carpets, soft furnishings and acoustic

Figure 6.3. The Northern facade of Montgomery school, showing the large windows for the classrooms.

panels to absorb sound waves, however, these would also mask the thermal mass. This is often an issue with modern office buildings in that all the thermal mass is located within the floor slabs but is insulated from heat transfer by carpet on one side and ceiling tiles (to improve acoustics and hide services) on the other. In Montgomery Primary School, since walls are used for student work, acoustic panels had to be located on the ceiling; however, these took the form of rafts rather than a continuous ceiling to allow air circulation and heat transfer to the concrete above. This required some computer simulations both to determine how much acoustic material had to be used to achieve the required reverberation times in the classrooms but also the size, layout, and separation of the individual rafts to allow the formation of convection currents within the classroom to allow heat from the pupils to be transferred into the floor/ceiling slab. The final result can be seen in figure 6.4; the rafts are able to hide the services and house the lighting fixtures while still allowing heat transfer between the air and all available thermal mass.

The majority of heat is generated in the classrooms; these are also the main occupied areas. We have already detailed what measures are taken to promote absorption of this heat into the concrete structure, but rather than waste heat that would otherwise be lost with exhaust air from the classrooms, the school employs several measures. The school, like many other Passivhaus', uses a mechanical ventilation system with heat recovery, with windows only opened in summer to control overheating. While heat exchangers are efficient (~85%) at removing sensible heat (and sometimes latent heat as well) from the exhaust air supply and transferring this to the incoming fresh air, it is more efficient to transfer as much heat to the structure as possible before air passes to the heat exchanger. Therefore, warm air is

Figure 6.4. Acoustic ceiling rafts in the sports hall suspended below the concrete ceiling to allow air circulation.

extracted from the back of the classrooms through openings (visible in figure 6.5) into the corridors, which are unheated spaces. The warm air passes along the corridors, and air from the ground floor is able to pass up through holes in the first floor corridor (figure 6.6) before being extracted at a high level. In this way, exhaust air has the maximum opportunity to give up its heat (both sensible and latent) to the thermal mass in the building structure before it reaches the heat exchanger. In order to prevent noise transmission between classrooms via the corridors, the ventilation holes are occluded with sound absorbing material, which can be seen in figure 6.6.

The holes in the first floor corridor also allow daylight from rooflights above to penetrate into the back of the classrooms, via 'vision panels', thereby improving daylighting uniformity in the classrooms. The ability to move heat around the building is particularly important not just for reducing heating bills for the school, but also in preventing overheating in the warmer months both now and in the future. The openable rooflights allow heat to escape from the school via buoyancy-driven stack ventilation, which does not require wind flow to be effective. The mechanical ventilation system allows the secure night-time purging of heat form the building without the opening of windows or rooflights. Removing the excess heat from the structure overnight refreshes the thermal mass, allowing it to store excess heat again the next day, reducing peak temperatures and preventing overheating. In this way the school can provide year-round thermal comfort even in the height of a 2050s summer under an aggressive estimate of climate change.

Montgomery Primary School is a useful example to provide in the final chapter of this book because it utilises many of the principles we have covered in earlier chapters, albeit applied in way suitable for British architecture. The site utilises water-permeable surfaces to reduce surface water flooding, and is raised to prevent fluvial flooding. Gutters and downpipes are also oversized to cope with more intense

Figure 6.5. View from a corridor into a classroom during construction, showing holes for the 'vision panel' and for the air extraction above.

Figure 6.6. Views along the corridors, ground floor on the left, and upper floor on the right. Warm air is extracted from the classrooms and passes along the corridors and up through holes in the first-floor corridor.

rainfall in the future. The concrete structure is not only strong to resist strong winds but also resistant to driving rain and flood damage. The highly insulated and airtight shell keeps heat out in summer and in in winter. Openings are shaded from direct solar radiation, and the vast amount of thermal mass regulates internal temperatures, while the provision for fan-assisted night-time purging allows the building to remain comfortable even during a prolonged heatwave event. While the building undoubtedly contains a huge amount of embodied carbon, it is highly efficient and is a net exporter of electricity to the national grid and does not overheat like its peers. The teaching staff have also noted an increase in student attainment since moving to the new school building which can linked back to the improved internal environmental conditions.

Montgomery was a purpose-designed new building, and its successes show that it is possible to be both energy efficient and resilient to a changing climate. When it comes to our existing housing stock, there are several key principles:

Shelter from the elements

Providing shelter is the fundamental purpose of our buildings. In order for this to continue, it may be necessary for us to provide shelter for our buildings from the extremes of weather. High winds and driving rain can cause damage to building fabric and possible structural failure. In exposed areas, it is a benefit to make buildings as aerodynamic as possible. Since wind speed increases with height, lower-lying buildings such as bungalows will be more resistant to high winds than two-storey or greater buildings. Thought should be given to the shape and orientation of the roof: Does the wind tend to come from a set direction? We should also reduce the area of the roof that faces directly into the wind, and choose an appropriate inclination so that wind pressure is reduced. For existing buildings, wind breaks can be deployed, such as the planting of trees a suitable distance away, or even the creation of grass banks upwind; these will disrupt wind flow and shelter the building from the high winds and driving rain.

Given the projected changes to rainfall intensity, the way we currently size our gutters based the amount of rainfall to fall in a minute, will mean that our current gutters even if kept clear will no longer be able to handle the volume of water running off our roofs. Overflowing gutters can lead to water running down the side of our buildings, which can lead to water seeping into our building and causing structural damage. Increasing the volume of the gutter or adding additional downpipes will help alleviate this problem. Alternatively, if the roof structure will take the additional weight, the use of vegetated roofs will slow rainfall run off and allow existing gutters to cope with run off (if any).

The increased intensity of rainfall may well require infrastructural upgrades to stormwater drainage. This can be alleviated at source (on site) through the incorporation of rain gardens and more permeable surfaces rather than concrete and impermeable paving. This will reduce surface water runoff and help prevent flash flooding of properties. Elements of water storage such as ponds or water butts are also a good idea to cope with the drier summer months.

Keep heat out

It is much easier to keep heat out of our buildings rather than to try and remove it once it is inside. Solar radiation can be up to 1000 W m^{-2} in the UK, which makes controlling the heat of the Sun the first line of defence when trying to prevent overheating, even for buildings with high internal gains such a schools and offices.

Reflecting solar radiation before it can be absorbed would seem to be the first step. Changing the emissivity of our buildings by simply painting them a lighter colour can help keep them cool. We only have to look at the many examples around the Mediterranean, such as Elba (figure 3.9), Cannes (figure 3.16), or Santorini (figure 3.19), to see how effective this can be.

The solar radiation that is not reflected is absorbed, raising the temperature of the absorbing surface according to its specific heat capacity, volume, and conductivity. More insulative materials will attain higher surface temperatures and will radiate more heat to their surroundings, with heat less able to pass through to the building interior. The most exposed part of any building to solar radiation is the roof, which is a large, unshaded, and often optimally inclined space, receiving the maximum amount of solar radiation per m^2. Therefore, increasing the amount of insulation in a roof will help limit how much heat from absorbed solar radiation can pass into the spaces below. Vegetated roofs are also of benefit here; in addition to their rainwater attenuation credentials, vegetation typically has a higher albedo than roofing materials, reflecting solar radiation back out to space, and the evapotranspirative cooling effects will actively cool the soil and building fabric below.

Any openings in our buildings such as windows and French doors need to be protected from direct solar radiation in the warmer months. East- and west-facing openings should be avoided if possible so that low-angle solar radiation at the beginning and end of the day cannot penetrate deep into the building, where it is unlikely to be able to escape readily. At lower latitudes, south- or north-facing openings are less at risk because the solar altitude is so great; however at mid-latitude locations, such as the UK, shading is required to prevent intense solar radiation entering our buildings. Balconies, roof overhangs, and fixed shades such as brise soleil are the obvious choices to shade openings and façades, but can be difficult to implement in existing buildings and can be considered unsightly. An easy-to-implement alternative is the use of a veranda or trellis to support plants, which will shade the opening in summer but allow passive solar heating in winter. Plants such as clematis are ideal for this since they are light, good climbing plants which can be easily supported. The use of vegetation will also provide some air filtration and some evaporative cooling of incoming air.

Finally, we should consider the materials we surround our buildings with. Figures 3.11 and 3.12 provide some indication that hard impermeable surfaces around buildings, particularly near openings, can lead to increased heating, even after the Sun has gone down. Ideally patios or driveways should be a little way away from the walls so that radiated heat does not enter our buildings. There is also a noted difference in the temperatures of water-permeable surfaces compared to impermeable ones, which combined with the benefits of reduced surface water runoff and increased percolation makes the choice of surface materials around our buildings fairly easy.

Remove excess heat

The removal of heat is key to providing thermal comfort and preventing summertime deaths in the future. Unfortunately, the highest temperatures and the most dangerous heatwave events are typified by a lack of wind to purge heat from our buildings. This means that even if we throw our windows open, we will struggle to eject the heat from our homes. Without resorting to a mechanical ventilation to purge the heat, we can instead utilise the density difference between air at different temperatures and altitudes. Buoyancy-driven ventilation is often not utilised due to concerns over security: we do not like to leave windows open on the ground floor. However, the inclusion of secure openings, whether vents or louvered windows, will allow the flow of air through our buildings at night to remove heat (or even during the day when we are not there).

Increasing access to thermal mass within our homes can be beneficial in reducing peak temperatures. Replacing wooden laminate flooring and carpet with ceramic or stone flooring will increase the amount of available thermal mass and also allow for conduction of heat into the screed/concrete below. Perhaps while not desirable, the replacement of dry-lined partitions with brick or block ones with gypsum plasterwork will provide additional thermal mass to help regulate temperatures. For office buildings, simply removing ceiling tiles around the perimeter of an office or replacing them with egg crate ones will allow air to circulate in the void space above and transfer heat into the otherwise masked thermal mass of the floor/ceiling slab above.

6.2 Urban resilience

The ideas of urban resilience are principally an extension of the ideas for building resilience. The choice of materials across an urban area is a determining factor in how resilient it will be to climate change. Most cities are located along rivers, which are typically constrained as they pass through urban areas. Making an urban area more permeable, to increase ground water percolation and slow and reduce surface water runoff, will alleviate pressure on these rivers, reducing the likelihood that they will flood and cause widespread damage. In the Netherlands, the 'Room for the River' project involves moving or removing dikes to allow rivers room to flood, providing sacrificial areas for rivers to expand into. This is a similar principle to the flooding of the Somerset Levels and to a lesser extent the Exeter's flood relief channel, both mentioned in chapter 2, as they provide somewhere for excess water to go. Such measures however, are not always possible, so instead we need to consider reducing the volume of water in rivers by other measures to limit fluvial flooding. Lessons can be learned from the destruction of Boscastle in 2004. Clearing of upstream scrub land meant that water entered the river faster, increasing the water flow rate trying to pass through the village. If rural areas upstream of urban areas can be hydrodynamically rough to slow water down, then water flow rates can be reduced and the likelihood of fluvial flooding reduced. Flash flooding from surface water is also a real issue in urban areas, and is only likely to become more significant as rainfall intensities increase as a result of climate change. While permeable surfaces help reduce this and it is now in fact a policy of many local authorities to not permit the creation of impermeable driveways for example, the water still needs somewhere to go if the ground becomes saturated, or if storm drains are already full. The creation of areas designed to collect rainwater within the urban areas as part of a wider SuDS programme, such as swales and raingardens (whether grass or paved ones), will allow the retention of water for a short while so that river levels can drop and drains can empty.

If such measures are still insufficient, then it may well be necessary to build defences to keep the water at bay, such as raising the height of river banks and building walls. For some coastal villages such as Fowey in Cornwall, higher defences are the only real option to cope with rising sea levels and increasing storm surges. Whether these defences take the form of higher sea walls or tidal barrages, such as those in the Thames or outside of Venice, will depend on the individual circumstances of the urban area in question and available funds.

The urban heat island (UHI) is a product of the layout of the buildings in an urban area and the activities going on inside them. The increase in urban temperatures due to the heat island is equivalent to the most aggressive estimates of climate change over this century, and the potency of the heat island is greatest during extreme heatwave events further increasing the risk to human life. To make matters worse, the formation of a heat island leads to the accumulation of airborne pollutants at street level. Therefore, the reduction of the UHI can be considered a key urban resilience measure.

The two main ways to reduce the UHI are to remove heat from the urban area and to encourage vertical transport of air. Altering the albedo of our buildings, roads, and pavements, will help reduce solar heating, thus encouraging the reflection of shortwave solar radiation back out to space before it is absorbed. The geometry of most urban areas leads to the absorption and radiation and reabsorption of longwave thermal radiation by roads and buildings. Allowing heat-absorbing surfaces to shed this heat is key to reducing the UHI. Increasing the sky view factor of a street will aid in this. There is also much to be said for making cities greener and bluer. The Garden City movement devised by Ebenezer Howard in 1898 and realised in the cities of Letchworth and Welwyn Garden City was originally created as a way to overcoming the unhealthy conditions and overcrowding that existed in other cities at the time. This philosophy has now been adopted by other cities and towns not just for the psychological and physiological benefits, but also as a way to reduce the UHI. In 1967, the Prime Minster of Singapore announced the intention to turn the city-state into a garden city with abundant greenery and a clean environment to make life more pleasant for its people. This lead to over 55 000 trees being planted in the next three years. In the mid-1970s the creation of new parks became a major focus of this 'Garden City' vision. By 2014, the number of trees planted had increased to ~1.4 million and the number of parks had increased from 13 to 330 with an increase in the area of green space from 879 ha (~0.9 km^2) in 1975 to 9702 ha (~97 km^2) in 2014. The benefits of such widespread greening are not just associated with the evapotranspirative cooling of the atmosphere, but also with an increase in surface roughness and perhaps more importantly an increase in the diversity of the roughness, leading to turbulence and increased vertical transport of air. Greenspace also typically has a higher albedo than other surfaces in urban areas, and the evapotranspiration of water is an endothermic process; both reduce the amount of heat in an urban area. While the environmental physics behind the formation or disruption of a boundary layer over a city is complex and perhaps not fully understood, the incorporation of networks of green and blue space throughout an urban area will lead to an increase in vertical transport due to the fact that water vapour is less dense than air. Additionally, increasing turbulence through the incorporation of varied surface types, building shapes, and sizes will lead to a greater diffusion of heat and pollutants into the atmosphere above.

Imagine a city, that instead of being hard, grey, and black, is soft, green, and blue, with vegetation and water defining the form and character of the surrounding buildings. Such a city would permeable to and absorbing of intense rainfall, and be cooled by the evapotranspiration of lush vegetation. Extensive green and blue spaces would provide valuable environmental capital, enhanced biodiversity, and public amenities. The diverse urban surface would disrupt stable boundary layer formation, disperse air pollution, and reduce urban temperatures to keep its citizens safe and comfortable at work, in their homes and on the streets. This, perhaps is the city of the future!

Lightning Source UK Ltd.
Milton Keynes UK
UKHW05n2010260618
324811UK00003B/61/P

9 780750 311984